Advancing Global Bioethics

Volume 22

The book series Global Bioethics provides a forum for normative analysis of a vast range of important new issues in bioethics from a truly global perspective and with a cross-cultural approach. The issues covered by the series include among other things sponsorship of research and education, scientific misconduct and research integrity, exploitation of research participants in resource-poor settings, brain drain and migration of healthcare workers, organ trafficking and transplant tourism, indigenous medicine, biodiversity, commodification of human tissue, benefit sharing, bio-industry and food, malnutrition and hunger, human rights, and climate change.

Henrik Lerner

Ethics for One Health Approaches: A Roadmap for Future Directions

Springer

Henrik Lerner
Department of Health Care Sciences
Marie Cederschiöld University
Stockholm, Sweden

ISSN 2212-652X ISSN 2212-6538 (electronic)
Advancing Global Bioethics
ISBN 978-3-032-05238-4 ISBN 978-3-032-05239-1 (eBook)
https://doi.org/10.1007/978-3-032-05239-1

This Springer imprint is published by the registered company Springer Nature Switzerland AG
The registered company address is: Gewerbestrasse 11, 6330 Cham, Switzerland

Acknowledgements

I started studying One Health approaches in approximately 2010 and realized that I had been conducting One Health research on health definitions since 2002 without knowing that this field existed. This book on ethics has been a project for me for a couple of years, and several people have contributed to my understanding as well as to the development of my thoughts along the way. I am not able to mention all, but I will here acknowledge those that have been particularly valuable to me in this process.

First, Charlotte "Lotta" Berg. We shared a car on the way home from a research meeting and accidentally ended up having a talk on One Health. That was something I had studied briefly, but she invited me to my first One Health Sweden research conference to talk about comparative aspects of health definitions. Then we co-wrote two influential papers on One Health approaches, which were partly forerunners to the Chap. 2 of this book. She also agreed to co-chair a track on Ethics, One Health, and the SDGs at the ISRDS conference in Stockholm 2022.

The presenters at the track of the ISRDS conference in Stockholm 2022—Rebecca E. Nordquist, Zohar Lederman, Jared Keyel, and Patricia Mooney Nickel—brought a new perspective to my thoughts. Zohar, whom I first met in Melbourne at the One Health conference in 2016, was one of the first people who led me to realize that ethics was a rather pristine field in One Health. Rebecca influenced me to include the Epigraph in the first chapter, which talks about wicked problems. Patricia and Jared had a very stimulating talk on how to form an ethical dialogue.

Simon Rüegg, who shares a similar interest in theoretical matters to mine, has been a nice discussion partner, and together we have been sharing discussions on the importance of interdisciplinarity, the scientific foundations of One Health, and cultural aspects.

Craig Stephen, with whom I share an interest in how to define health, reminded me and drew my attention to justice matters and to how important it is to collaborate with Indigenous peoples. We met in person at the One Health Conference in Saskatoon in 2018, a conference where issues of injustice, gender structures, and colonial research were brought forward by the participants. I had worked several years earlier with the concept of traditional knowledge and Indigenous peoples' claims on research ethics, and after the conference I saw a much more obvious connection between One Health and these issues than I had before.

David Fraser, who kindly provided me with documents that improved my section on his theory of animal ethics and offered suggestions on how to understand One Welfare.

Vilhjalmur Árnason, with whom I had several days of discussion on dialogue ethics during two adjacent conferences in Riga where we both spoke, helped improve and strengthen my suggestion for dialogue ethics in Chap. 7. The conferences were hosted by the Nordic-Baltic Network for Philosophy of Medicine and the European Society for Philosophy of Medicine and Healthcare, and I thank the organizers for letting me speak about parts of Chap. 7, as well as the participants for giving fruitful thoughts on the matter.

Several people drew my attention to important texts and gave suggestions on the choice of cases, as well as offered good reflections on early drafts. I appreciate all these suggestions, even though some are not included in the book:

Lars Andersson, Glenn Cosquer, Staffan Dryselius, Sara Eklöf, Stefan Eriksson, Ingrid Hellström, Anna T. Höglund, Peter Lundberg, Filip Maric, Jo Middleton, Kajsa Norbäck, Malin Olsson, Pär Segerdahl, Amanda Thorell, Olle Torpman, Julia Whitfield, and Jane Österlind.

The staff at Marie Cederschiöld University library, who helped me obtain several of the references through interlibrary loans.

All the staff at Springer Verlag. This book would never have been finished if they hadn't believed in the idea. Special thanks to Floor Osting, who was my first contact and has been very helpful through the process. Travis Holloway, who helped me finalize the project. The editors of the book series Advancing Global Bioethics, Henk ten Have and Bert Gordijn, who gave me the opportunity to be a part of it. Kanchan Latambale for help with additional machine-generated literature searches. Arun Santhosh Kannan, Immaculate Jayanthi, and Radha Lakshmanan for all technical support.

Finally, this book project would never have been realized without the support of Liza Hagberg Lerner. Through daily discussions, she has improved and problematized my thoughts in a fruitful way. She has meticulously read every chapter and provided many valuable comments that have improved my writing.

Competing Interests The author has no competing interests to declare that are relevant to the content of this manuscript.

Contents

Chapter 1
Introduction and Scope of the Book

> *Infectious zoonoses emerge from complex interactions among social and ecological systems. Understanding this complexity requires the accommodation of multiple, often conflicting, perspectives and narratives, rooted in different value systems and temporal-spatial scales. Therefore, to be adaptive, successful and sustainable, One Health approaches necessarily entail conflicts among observers, practitioners and scholars. Nevertheless, these integrative approaches have, both implicitly and explicitly, tended to marginalize some perspectives and prioritize others, resulting in a kind of technocratic tyranny. An important function of One Health approaches should be to facilitate and manage those conflicts rather than to impose solutions. (Waltner-Toews 2017, 1)*

Abstract The aim of this book is to try to find frameworks for how to solve ethical dilemmas within One Health approaches. This first chapter introduces One Health approaches, their scope and history arguing that ethics have been neglected within these. A brief introduction in the field of ethics is made with the aim of distinguishing some important concepts that will appear in forthcoming chapters. Four ethical trends are introduced and briefly mentioned how they influence this work. The trends are expanding ethics, questioning the individualistic take, pluralism, and constructivism. Then there follows a presentation of the structure of the book with the different topics covered. Finally, three ethical cases will be presented which will follow the reader through Chaps. 4 to 7 as an analytical tool to understand the differences in the ethical viewpoints presented and discussed. These real-world dilemmas cover important aspects of One Health approaches.

Keywords One Health approaches · Ethics · Values · Individualism · Ethical cases · Pluralism · Wicked problems

H. Lerner, *Ethics for One Health Approaches: A Roadmap for Future Directions*, Advancing Global Bioethics 22,
https://doi.org/10.1007/978-3-032-05239-1_1

1.1 One Health Ethics in One Health Approaches

This book will introduce One Health ethics, a fairly new topic in bioethics, which has quite recently been starting its winding road as a scientific field. The aim of this book is to find frameworks for how to solve ethical dilemmas within One Health approaches. Starting with a description of the different One Health approaches and suggestions for ethical frameworks already proposed within these approaches. I will contribute with a theoretical in-depth analysis of these possible ethical roads as well as an application of each framework on three cases to see whether they are useful in real-world scenarios. The result of this will be a roadmap for future research where I try to sort out different, fruitful roads, while clarifying why we should avoid some other less fruitful. In this chapter, I will clarify some of my original standpoints and outline the rest of the book.

Shortly put, One Health ethics is ethics in One Health approaches. One Health approaches are in its widest sense focusing on and promoting health for all organisms including the systemic level such as ecosystems. They are inter- or transdisciplinary. What I will try to show in this book is that the ethics needed within these approaches might be one of the widest versions of bioethics. The approaches have arisen as answers to complex problems[1] such as the one mentioned in the epigraph by David Waltner-Toews. Another example is the discussion by Johan Rockström and Owen Gaffney (2021) on planetary boundaries that our global lifestyle drives the world to tipping points which might be nonreversible. They claim that we need to reverse the process before it is too late. In One Health approaches these complex problems all refer to different kinds of health issues. And health tends to be seen holistically. For example, to study an infectious disease system (such as a zoonosis[2]), one cannot reduce it down to its parts; one needs a holistic approach (Garira 2017).

The term "One Health" are in this book used in two ways, in line with what has been suggested earlier (Lerner and Zinsstag 2021). This resembles the present usage in scientific and grey literature where One Health sometimes are referring to several approaches and sometimes only to one particular approach. To avoid confusion, I use 'One Health approaches' when I refer to all approaches, such as One Health, EcoHealth, Planetary Health, and One Welfare (which will be described in-depth in Chap. 2). When I use 'One Health' I refer to a specific approach, the One Health approach, which originated from the One Medicine approach. Through this book I will consequently use "One Health approaches" for the former and "One Health" for the latter.[3]

---

[1] Among these are wicked problems, which are a special sort of problems that has its own theory (see further in Waltner-Toews (2017) for an introduction).

[2] Diseases that are transferred between humans and animals.

[3] I also choose to use capital letters for these terms throughout this book indicating established names for these approaches. In present scientific literature capitals are sometimes used and sometimes not.

The term 'One Health' is consisting of two words, which both characterize two crucial features for all approaches. 'One' stands for inter- or transdisciplinary which are core ways of doing research within the approaches and 'health' indicates the core value of all approaches (Lerner and Zinsstag 2021). 'Health' is seen wider than humans and could be attributed to humans, animals, plants and/or ecosystems (Lerner and Berg 2015). Health is concerning interspecies aspects of health and could be defined in reductionistic as well as holistic definitions (Lerner 2020). Another way to explain One Health approaches shortly and popular is proposed by Evans and Leighton (2014, 415) using health as an acronym for "Humans Ecosystems Animals Living Together Harmoniously".

Living together harmoniously, which is also a core aspect of the three pillar model of sustainability,[4] calls for ethics. Most of the United Nations' *Sustainable Development Goals* (abbr. SDGs) could therefore be tied to One Health approaches (Queenan et al. 2017; de Paula 2021).[5] Ethics for One Health approaches might be the widest version of bioethics where one needs to consider almost all aspects of life if it concerns health. This will not be an easy task, but this book will be both an introduction to and an in-depth analysis of this subject. In a way this book will try to answer the call from Waltner-Toews in the epigraph although he didn't specify it in ethical terms. The claim for an ethical framework for One Health approaches has been around since the mid-2010s (see Chap. 3) and some authors (Destoumieux-Garzón et al. 2018) argue for a broad inclusion not only of human bioethics but also of animal and environmental ethics. Nonetheless, an in-depth discussion is still lacking. This book will be a roadmap towards this.

1.1.1 Overview of One Health Approaches

During the late twenty-first century, an understanding of health as complex and interconnected between humans, animals and ecosystems evolved in several areas of science. The One Health approaches evolved from somewhat different directions. They differ somewhat in core aspects such as contributing sciences and their influence, their view of contributing knowledge (for example if also Indigenous knowledge is included besides western scientific), their emphasis on which health that is important, and the goals for the approach.

Among the One Health approaches, the one with the oldest roots is One Health, stemming back to the nineteenth century and the theoretical works of One Medicine

[4] There are two common interpretations of sustainability, the Brundtland definition and the three-pillar model. The Brundtland definition is anthropocentric with a focus on sustainability for humans, while the three-pillar model includes environmental, social, and economic sustainability.

[5] There has been critique that the SDGs emphasize human sustainability more than the three-pillar model does. For example, Keeling et al. (2019) tested the SDGs with regard to animal welfare and proposed that their method also could be used for other issues that are not yet part of the main focus of the SDGs.

by Rudolf Virchow (1821–1902) combining medical and veterinary science (Lerner 2012, 2013). The starting point for the shift from One Medicine to One Health has been set to 2004, when there was a conference where the 12 Manhattan Principles linking humans, animals and ecosystem was proposed (Cook et al. 2004). One Health combines human medicine, veterinary medicine, and ecology, although emphasis is on human and veterinary medicine.

In 2004, the Ecohealth journal established the field of EcoHealth (Saint-Charles et al. 2014), which merged from the approaches of ecosystem health, global change and health, and Conservation Medicine.

The roots of Planetary Health lie in the 1970s and 1990s but it was established as an approach in mid 2010s (Horton et al. 2014; Whitmee et al. 2015; Almada et al. 2017; Prescott and Logan 2019). The approach has been favoured by the Rockefeller Foundation and the scientific journal *Lancet* through the Rockefeller Foundation-Lancet Commission on Planetary Health.

One Welfare arose in the 2010s (Anonymous 2012; Colonius and Earley 2013; Mills and Hall 2014; García Pinillos et al. 2016; García Pinillos 2018), with an origin from One Health trying to combine animal, human, and social welfare.

These four approaches will be further explained in Chap. 2.

1.2 Ethics

I call my book *Ethics for One Health Approaches: A Roadmap for Future Directions*. The idea of a roadmap[6] is that there could be alternative roads, some better, some worse. I am therefore not aiming for *the* theory of ethics. I would like to explore which ethical ways are fruitful in themselves or in combination with others. I think finding *the* theory is a utopia. Since the first ethical texts were written, we have quarreled about the right version without ever finding a consensus globally. With this said, I strongly believe that we can do better in ethics and develop our ethical theories so that they better do their work. This book will also try to do this development. First, I need to say something introductory on the scientific field of ethics and my view on ethics.

There are several ethical theories that could be relevant for One Health approaches. How could one distinguish between them? This could be done in several ways, and I will give a short introduction here so that the reader will get an overview of the field.

First, in ethical curriculum one often teaches and distinguishes between five ethical (mainly) anthropocentric theories, namely utilitarianism, deontology,

[6] Roadmap as a concept has been used earlier in the One Health literature. Nicole de Paula (2021) has used it as giving the direction for Planetary Health in the future. Zohar Lederman (2018) has used it in a pragmatic way to avoid defending his claims of theory in his ethical thesis on One Health and culling (see Chap. 3). My intention with the concept differs from his. I return to the concept of a roadmap in ethics in Chap. 8.

contractual ethics, virtue ethics, and feminist ethics (Shafer Landau 2012). These are specific normative theories presented at the individual level.

Second, at the societal level, one can in line with Will Kymlicka (2002) distinguish between at least 8 normative positions; utilitarianism, liberal equality, libertarianism, Marxism, communitarianism, citizenship theory, multiculturalism, and feminism.

Third, one can distinguish between theories that are primarily based on either *ethical principlism*, where one applies principles or rules to an ethical dilemma, or *ethical proceduralism*, creating a procedure that will be ethical. An ethical theory could be seen as one of the richest and most overarching level. The theory structures whether ethics is mainly procedural- or principle-based, which procedures and principles that are core, and at the most particular level, how these relate to ethical concepts and values. So, what is the difference between concept, principle, and theory? At the lowest level, an ethical concept, for example equality, could be just mentioned as something valuable or sought for. Nonetheless, one gets no further information on how to facilitate equality. An ethical principle might be a formulation that could work as a guide, such as how to redistribute unequal resources. If placed in a declaration or used as a rule-of-thumb it helps people to decide in ethical matters but leave out how it might relate to other ethical aspects. An ethical theory is the richest, placing an ethical concept or principle in relation to other ethical concepts, principles, or procedures. One example here could be a feminist theory on gender equality.[7]

A fourth way to distinguish is to use value theory (normally placed within the field of meta-ethics). The common distinction between ethical values is between intrinsic and instrumental values. Intrinsic values are something valued that is good in itself. Instrumental value is something good that helps to reach an intrinsic value. A third kind of value, which will be introduced and discussed in Chap. 6, is relational value which is used in species and ecosystem preservation. Theories could be arranged with regard to how one values individuals and how one demarcates who or which will have moral status. Here, one differs between *anthropocentrism*, *zoocentrism*, *biocentrism*, and *ecocentrism*. In anthropocentrism, human individuals are the only ones valued. In the strong version, only humans have moral status, but in the weak version, also individuals of other species might count but always less than humans. Zoocentric theories are often called animal ethics and value both human and animal individuals (depending on theory there will be a difference in which animals that are considered to have moral status). Biocentric and ecocentric theories mainly fall within environmental ethics and include also other organism groups than animals as well as systems like ecosystems. In environmental ethics, values are often analyzed, and one way is to compare different kinds of values such as ethical, economic, aesthetic, and even others. Biocentric theories value all organisms that have life (also including for example plants). Ecocentric theories mainly place value

[7] See Kymlicka (2002, 1–9) for an interesting discussion on the relation of concepts, principles and theories and how one can use them to compare different viewpoints, which has inspired me in thinking on concepts, principles, and theories.

in a more aggregated order than individuals, either in the species (for example genetically similar populations as in biodiversity) or in ecosystems (for example well-functioning processes that involve individuals). In both these, like biocentrism, all organism forms might have moral status.

Some of these terms mentioned will not be in focus throughout the book while others will be crucial. In this book I will mainly use the third and fourth distinction but through the text some of the other terms from the first and second distinctions will appear.

1.2.1 Four Trends in Ethics

To do better in ethics, one has also to be aware of recent trends in ethics to grasp the latest developments and consider if these are useful or valuable. As I have understood the development of ethics in philosophy during the twentieth century up to now, there have been at least four trends one has to be aware of if one is aiming for a modern approach to ethics. These are

1. Expanding ethics
2. Questioning the individualistic take
3. Pluralism
4. Constructivism

1. Since approximately the 1940s there have been claims for *expanding ethics* to include also animals and ecosystems besides humans.[8] A core discussion throughout the book will be on *who or which counts* within the sphere of ethics or morals. There are several terms for those entities that we should consider in ethical analysis, for example moral status, moral standing, or moral considerability (Beauchamp and Childress 2019, 65). I will use the term *moral status* throughout the book for these matters.

For me the moral community (those that have moral status) comprises both moral agents, who could be held responsible for their ethical choices, and moral objects (or patients), which are all other entities or organisms[9] that we consider in ethical consideration. For moral objects I follow Beauchamp and Childress (2019, 67) distinction of direct moral object (for example an animal) and indirect moral

[8] Earlier accounts exist especially for animals, see for example Thomas (1984). Aldo Leopold (1887–1948) is in environmental ethics often referred to as the first to talk about an expanding circle including wider and wider areas of moral objects. His central writings appeared in the 1930–40s.

[9] In this book I follow my earlier assumptions (Lerner 2020) on organisms, excluding viruses and prions but including all life forms, such as all kinds of bacteria, plants, fungi, and animals both unicellular and multicellular. These are sometimes gathered in seven kingdoms (see Lerner (2020), for further references to this debate).

object (for example inanimate things). The demarcation of moral agents as well as direct and indirect objects differ between ethical theories.

For moral status, *value theory* will also be crucial. How should moral objects be valued and will the value be absolute or graded in comparison to moral agents? Is there only one value to consider or are there several? These are questions that I will try to address through the book.

2. During the twentieth century a strong emphasis on the *individual* was established especially in western healthcare ethics. At the end of the twentieth century and start of the twenty-first century this has been contested by feminist ethics,[10] dialogue ethics[11] and non-western ethics.[12] One assumption in dialogue ethics, that questions the individualistic take, is that one cannot decide ethical issues by oneself, one needs to get together in order to grasp all perspectives and alternatives (Benhabib 1996). Another critique from Indigenous and feminist ethics is that individuals as interconnected with the community and ecosystems (Szende 2022).

3. Claims for *pluralistic* ethical theories have grown due to the inherent problems of simplified ethical theories focusing on one core aspect (such as utilitarianism or deontology at the individual level). They seem to be insufficient to cover the entire field of ethical situations, and pluralism seems to be a possible option. Besides those ethicists trying to bridge ethical theories within the western ethical sphere, also non-western ethics has come to the fore.

Which will become evident in Chaps. 2 and 3, this book needs to discuss possible ways to bridge a gap between anthropocentric, zoocentric, biocentric and ecocentric ethical theories. In the first doctoral thesis on ethics and One Health, Zohar Lederman (2018) argued that there is an ethical imperative in the definitions of One Health and that imperative will force ethics to consider more than anthropocentric public or global health ethics, but rather the entire spectrum of ethical positions. This book will therefore consider ethics for humans, animals, plants and ecosystems, and try to bridge ethics about individuals, populations, and systems. Aware that this has not yet been done, my roadmap in this book will be one step towards such a bridge rather than a final solution.

4. Connected to pluralism, constructivism has gained more acceptance. Theories put forward nowadays argues rather for an acceptance of arguments that everyone can agree on through deliberation as fair or proper, rather than a reference to an objective moral standard all of us must realize in our minds. My take is that we need a well-considered *constructivist* position, an ethical framework (or a few ethical frameworks) we can agree on that will be useful within the One Health approaches.

---

[10] For one example of the feminist discussion about traditional views on individual autonomy and the critique from feminism see Friedman (1997).

[11] Also labelled communicative ethics, discourse ethics, and dialectical ethics depending on differences between the theories in the same strand. I here use dialogue ethics as the term to refer to this strand in ethics which includes scholars such as Karl-Otto Apel, Jürgen Habermas, and Seyla Benhabib.

[12] For one example, see the discussion on the postcolonial African concept of Ubuntu (Eze 2023).

In my book I will use empirical material gathered from scientific texts within the One Health approaches as an initial starting point to set the arena for ethics. This will be an analysis of how these approaches is defined as well as thoughts on ethics already present in the field. Here I must meet a possible objection to my method. Will I commit an is-ought fallacy[13] while using empirical material from the research within One Health approaches to say something about the normative aspect of what One Health approaches need for ethics? To avoid this problem, I will follow Jonathan Glover's (2001, 6) position in his book *Humanity*. He indicates three ways of dealing with empirics in ethics. The most theoretical way is when moral principles are applied to empirical matters without changing the ethical theory. Then empirics will not influence the theory. As a middle ground, one could use ethical principles tentatively and modify them by our reasoning in the empirical dilemmas. This is consistent with John Rawls' Reflective Equilibrium, which is one of the most established ethical decision models. The third way is that one could study empirical matters to revise ethical principles or beliefs. This is a bottom-up-version that starts with empirical matters, which is the way Glover justifies for his own work. This is also a common approach in applied ethics. In this book I start with the bottom up-approach to get the scope of ethics needed, suggested ethical frameworks for issues as well as value theory within One Health approaches. Further on in the book the two other ways will be used to justify ethical frameworks that could be useful to try to avoid the is-ought fallacy. Here, One Health approaches must learn from theoretical ethics and middle ground theories to deepen ethical discussions as well as the solutions. Through the rest of the book, I will use the term ethical framework for those theories, principles, procedures or other structured ways of making an ethical decision I consider.

1.3 Structure of the Book

This book will analyze fruitful and useful ethical frameworks that could be applied within the One Health approaches. Without restrictions this task would be enormous if I would analyze all ethical theories proposed within ethics. Therefore, in this first part of the book, I will start my work in an empirical analysis of scientific literature on One Health approaches to analyze the width of bioethics needed (Chap. 2). The questions answered will consider the scope of the approaches and some relate more to meta-ethics. Such as: Which organisms seem to be considered, all or some?

[13] The is-ought fallacy is a common discussion when empirical ethics is discussed. The fallacy that one cannot derive a normative conclusion from premises that are empirical is attributed to David Hume, the so-called Hume's law (Blackburn 1996, for an example of the debate on the law, see Salwén 2003).

Where is focus put, on individuals, populations or systems such as ecosystems? Which stakeholders?[14] Which values are core within the approaches?

The next step (Chap. 3) is to summarize and analyze ethical attempts already being made within the One Health approaches. Gathered are both suggestions on which ethics to use as well as those calls for ethics which specify aspects that a One Health ethic needs to consider. This is one of the first comprehensive reviews of the state of the art of One Health ethics. This empirical study will give an input to the second part of the book where I will analyze in depth with some of the fruitful frameworks to find a roadmap of ethics for the One Health approaches. For both Chaps. 2 and 3 I have included what I find relevant documents up to June 2023.

The second part of the book (Chaps. 4–7) will be a structured analysis of different already proposed frameworks with additions that I find fruitful. I start in Chap. 4 with proposals based on anthropocentric ethics, use Chap. 5 for zooocentric ethics, Chap. 6 for bio- and ecocentric ethics, and end in Chap. 7 with pluralistic ethics. These frameworks will be analyzed in two steps. First, normatively to see their strengths and shortcomings. Second, by applying each of them to three real world cases, which I introduce in the next section (Sect. 1.4). This application is performed to understand how the different frameworks analyze the same cases, clarifying some of the limits of these frameworks when they are applied to real world cases, and what ethical implications will be in the different cases. In Chap. 6, I will also consider meta-ethics and analyze relational value, proposed by the Intergovernmental Platform on Biodiversity and Ecosystem Services (IPBES) as a third kind of ethical value, besides intrinsic and instrumental. I will try to see if this third category is proper and helpful in the discussion of ethics in general and One Health ethics in particular.

Finally in the third part of the book (Chap. 8), I will end in a conclusion about what is needed for the future discussion of One Health ethics. I will propose a roadmap that will suggest promising ways, sorting out some dead ends and indicate some obstacles to avoid along the road forward.

1.4 Ethical Cases

As already mentioned in the last section I will work throughout Chaps. 4–7 with three real world cases which I believe grasp several important ethical issues. Obviously, I cannot cover all possible ethical issues in these cases, so I have emphasized some crucial aspects in them to do the work. One guiding question is, could these evaluated ethical frameworks give sufficient insight into these cases or are the application of the cases still hard to perform?

---

[14] I use a wide meaning of the term "stakeholder" throughout the book which will include all that have some kind of interest in the issue discussed.

In the first case, I turn to the transmission of infectious diseases in what I call the Bats case. Among all possible ethical aspects, this case primarily deals with our relations with wild animals in their environment and how we consider conflicts when we attribute values. Should we attribute value to individuals, species, or ecosystems? Also, this case focus on possible conflicts between different value systems based on knowledge traditions such as western scientific and indigenous communities.

Then I present the Animal Assisted Therapy case. This case is about animals (i.e. domestic animals or humans in our care) in human health care when they are part of Animal Assisted Therapy. Focus is on therapy, relations between animals and humans, and mental health. Among all possible ethical aspects this case primarily deals with how we understand and work with relations-based ethics, animals as moral objects or potential moral agents, and whether some aspects could be tied to certain species (species-specific ethics).

Finally, the Oceans case is about oceans, climate and ecosystem change. Among all possible ethical aspects, this case primarily deals with how we value non-living parts of the ecosystem, pollution, eradication of resources shared by many nations, and environmental justice.

1.4.1 Bats Case

The philosophical question "what is it like to be a bat" proposed by Thomas Nagel (1974), has some relevance also to One Health approaches. Nagel had the focus on how to understand consciousness when he asked the question. In One Health approaches there is another ethical angle on the question. In the aim of finding new dangerous diseases before the outbreak, much emphasis has been on bats as an animal species group (Klimpel and Mehlhorn 2014). Many diseases are linked to bats, for example Corona viruses and Ebola (Lu et al. 2021) and the transmission cycle could be between insects—mosquitos, sandflies, and bedbugs—bats and humans (Melaun et al. 2014). This has also developed so far that bats could be pinpointed as a specifically dangerous species group for humans. For example, the project PREDICT, which has been a part of USAID's Emerging Pandemic Threats (EPT) Program, has for example published a moderated picture book that could be used as an "community engagement tool" named *Living Safely with Bats* (Francisco et al. 2018), indicating that the species group is dangerous to humans.[15]

Bats as a group is important in ecosystems and have several important roles such as pollination, seed dispersal, arthropod suppression and nutrient cycling (Kunz et al. 2011; Castillo-Figueroa 2020). Nutrient cycling is important for the local ecosystem but is also favourable for human agriculture. There are existing guidelines

[15] On the webpage, they state it as a "community engagement tool". The book itself is labelled as a moderated picture book.

on how to deal with guano harvesting in bat caves to minimize destruction of habitats, the risk of declining populations of bats and the risk of unsanctify religious places, especially for Indigenous peoples (IUCN SSC 2014). For humans, bats are used in some places as food and sometimes in limited medical aspects (Kunz et al. 2011).

Considering these two aspects, risk of catching diseases and bats as important for ecosystem services, brings the ethical issues in the case forefront. Central issues in this case are how to look at ecosystems and specific species groups. One core conflict could be between species conservation and disease transmission. What kind of role do diseases have in ecosystems and will the risk of transmission influence how we consider the survival of a species? Preserving biodiversity and disease eradication might be in conflict, which has previously been discussed (Jenkins et al. 2015; Middleton et al. 2020; Lu et al. 2021). One implication might be that promoting all the sustainability goals need a deliberation of which goal to promote. Here, SDG3 Good health and well-being mightconflict with SDG15 Life on land.[16] In this conflict, one argument could also be that healthy ecosystems should include, for example parasites (Jenkins et al. 2015). Other central issues include whether these animals are devalued due to the conflict, compared to similar animals that do not pose a risk of disease transmission, and how to respect traditional livelihoods in these habitats.

1.4.2 Animal Assisted Therapy Case

The Animal Assisted Therapy (AAT) case is about finding an interspecies ethics to evaluate animal-assisted interventions. These interventions have been used as examples of fields which is more holistic than just treating biological infections (Mills and Hall 2014; García Pinillos 2018), mainly from the One Welfare approach (see Chap. 2 for an elaboration on the approach).

Animal-assisted interventions is when animals provide care for humans. These interventions are divided in four branches: animal assisted therapy, animal assisted interventions, animal assisted learning, and animal assisted coaching/counselling (IAHAIO 2018). The reason for including this case is that in animal assisted interventions almost all aspects of human health could be included. Therefore, this is a rich case considering mental health, social health, and existential health besides

[16]The problem of conflicts between SDGs is a potential growing research area. For some other examples where this has been mentioned see Lerner et al. (2011, conflict between SDG13 Climate action and SDG15 Life on land), and Hickel (2019), arguing that SDG8 Decent work and economic growth is in conflict with goals that focus on humans as in "harmony with nature" (SDG6 Clean water and sanitation, SDG12 Responsible consumption and production, SDG13, SDG14 Life below water, and SDG15)).

physical health. Animal assisted therapy is the branch that most obviously deals with all these aspects of health. Therapy also has a goal in changing an individual.

The easiest setting for an animal assisted therapy is an animal, a treater, and a client, but depending on the goal of the therapy more than one animal, treater, or client might be included. The core in this therapy form is relations rather than environment, and the more complicated setting the more relations to consider.

Domestic animals such as birds, rabbits, cats, dogs, and horses are involved (IAHAIO 2018) where the horse has the widest therapy use due to its nature of being an animal both to take care of and to exercise with. The other animals could also be a part of human exercise, but horses provide a specific opportunity to be used in physiotherapy due to the possibility to ride on horses (Ekholm Fry 2021). Horse assisted therapy will therefore be the focus in this case. Most horse assisted therapy is performed on farms or riding houses so an environmental aspect of being outdoors is also possible in this case.

Even if there is strong emphasis on human care in these interventions (which is reflected in few studies on the welfare of animals in AAT (Ng et al. 2018)), one must not forget the animal which contributes to this. Some consider even the animal involved as an actor or agent, labelling it as a co-therapist (Levinson 1962; Forsling 2020). Therefore, finding an interspecies ethics that could analyze the human, the horse and the relations between these two species is crucial. It also gives a general knowledge of how to treat animals in human care.

Good hygiene standards and regular veterinary visits for the animals involved are important for good practice (IAHAIO 2018). One conflict in this case could be that while promoting several aspects of health, AAT might also facilitate the spread of diseases, for example an animal could be a vector for AMR spread between healthcare facilities.

1.4.3 Oceans Case

There has been less interest in the marine ecosystems than in the terrestrial ecosystems. One reason could be that we, as humans, find it easier to understand ecosystems above water because we live in them permanently. Still, the climate crisis has put emphasis on these ecosystems. According to de Paula (2021, 62), oceans could be considered the "lungs of the earth" rather than the tropical forests on land and changes in these ecosystems might have dramatic consequences. Melting ice due to higher temperatures increases the amount of water in the oceans. Sea level rising risks flooding local communities' livelihoods. Shorelines may move due to rising sea levels. Species extinction occurs in coral reefs due to warming, acidification and deoxygenation (Hughes et al. 2020). In 2008, dead zones in the water around coasts globally occurred in a total area of 245,000 square kilometers (Diaz and Rosenberg 2008). Fish populations are being overfished (Coll et al. 2008; Sumaila and Tai 2020) to get healthier and climate-friendly food and to avoid red meat. Another issue is the overuse of plastics which ends up as waste in the oceans and when plastics disintegrate, they form microplastics that when entering animal and human bodies might risk the health of the individual (Prata et al. 2021).

When climate warming warms the ocean, it creates more ocean acidification, and ocean deoxygeneation. This in turn decreases ecosystem services, food and nutrition security, biomedical and pharmaceutical compounds, and livelihoods as well as increases pathogenic organisms and disaster risk (Talukder et al. 2022). Weather changes and changes of currents in the oceans due to climate change will alter terrestrial ecosystems either to be more arid or more wet due to rains.

Among all possible ethical aspects this case primarily deals with how we consider the value of non-living parts of the ecosystem, pollution, eradication of resources (overfishing to reach climate goals), resources that are shared by many nations, ecosystems, and environmental justice.

1.5 Summary

One Health approaches might be the ultimate test to what the widest version of bioethics might be able to do. Ethics for these approaches needs to be able to bridge some of the perspectives (anthropocentrism, zoocentrism, biocentrism and ecocentrism) that today is separated. Which will become evident, one cannot fulfil the ethical imperative within the One Health approaches (which will be clearer to the reader after Chaps. 2 and 3). Next chapter will analyze similarities and differences in the main One Health approaches and give insights in what kind of ethics needed.

References

Almada, Amalia A., Christopher D. Golden, Steven A. Osofsky, and Samuel S. Myers. 2017. A case for planetary health/geohealth. *GeoHealth* 1:75–78.

Anonymous. 2012. Thinking beyond one health. *Veterinary Record*. https://doi.org/10.1136/vr.e3116

Beauchamp, Tom L., and James F. Childress. 2019. *Principles of biomedical ethics*. 8th ed. Oxford and New York: Oxford University Press.

Benhabib, Seyla. 1996. Toward a deliberative model of democratic legitimacy. In *Democracy and difference: Contesting the boundaries of the political*, ed. Seyla Benhabib, 67–94. Princeton, New Jersey: Princeton University Press.

Blackburn, Simon. 1996. *Oxford-dictionary of philosophy*. Oxford and New York: Oxford University Press.

Castillo-Figueroa, Dennis. 2020. Why bats matters: A critical assessment of bat-mediated ecological processes in the neotropics. *European Journal of Ecology* 6 (1): 77–101.

Coll, Marta, Simone Libralato, Sergi Tudela, Isabel Palomera, and Fabio Pranovi. 2008. Ecosystem overfishing in the ocean. *PLoS One* 3 (12): e3881.

Colonius, Tristan J., and Rosemary W. Earley. 2013. One welfare: A call to develop a broader framework of thought and action. *JAVMA* 242 (3): 309–310.

Cook, Robert A., William B. Karesh, and Steven A. Osofsky. 2004. *The manhattan principles on "one world, one health"*. http://www.wcs-ahead.org/manhattan_principles.html

de Paula, Nicole. 2021. *Breaking the silos for planetary health: A roadmap for a resilient post-pandemic world.* Singapore: Palgrave Macmillan.

Destoumieux-Garzón, Delphine, Patrick Mavingui, Gilles Boetsch, Jérome Boissier, Frédéric Darriet, Priscilla Duboz, Clémentine Fritsch, Patrick Giradoux, Frédérique Le Roux, Serge Morand, Christine Paillard, Dominique Pontier, Cédric Sueur, and Yann Voituron. 2018. The one health concept: 10 years old and a long road ahead. *Frontiers in Veterinary Science* 5:14.

Diaz, Robert J., and Rutger Rosenberg. 2008. Spreading dead zones and consequences for marine ecosystems. *Science* 321:926–929.

Ekholm Fry, Nina. 2021. Welfare considerations for horses in therapy and education services. In *The welfare of animals in animal-assisted interventions: Foundations and best practice methods*, ed. Jose M. Peralta and Aubrey H. Fine, 219–242. Cham: Springer.

Evans, B. R., and F. A. Leighton. 2014. A history of one health. *Revue Scientifique et Technique de l OIE* 33 (2): 413–420.

Eze, Michael Onyebuchi. 2023. Ubuntu/Botho. In *Key concepts in world philosophies*, ed. Sarah Flavel and Chiara Robbiano, 283–291. London: Bloomsbury Academic.

Forsling, Sven. 2020. Den terapeutiska relationen: hästen, flickan och terapeuten. In *Hästen och den mänskliga hälsan*, ed. Gunilla Silfverberg and Henrik Lerner, 88–130. Stockholm: Appell Förlag. [In Swedish].

Francisco, Leilani, Ava Sullivan, Jonathan Goley, and Stephanie Martinez. 2018. *Living safely with bats*. https://ohi.vetmed.ucdavis.edu/programs-projects/predict-project/publications. Retrieved 10 October 2022.

Friedman, Marilyn. 1997. Autonomy and social relationships rethinking the feminist critique. In *Feminists rethink the self*, ed. Diana T. Meyers, 40–61. Boulder, Colorado and Oxford: WestviewPress.

García Pinillos, Rebecca. 2018. *One welfare: A framework to improve animal welfare and human well-being*. Wallingford: Cabi.

García Pinillos, Rebeca, M. Appleby, X. Manteca, F. Scott-Park, C. Smith, and A. Velarde. 2016. One welfare – a platform for improving human and animal welfare. *Veterinary Record*. https://doi.org/10.1136/vr.i5470

Garira, Winston. 2017. A complete categorization of multiscalc models of infectious disease systems. *Journal of Biological Dynamics* 11 (1): 378–435.

Glover, Jonathan. 2001. *Humanity: A moral history of the twentieth century*. London: Pimlico.

Hickel, Jason. 2019. The contradiction of the sustainable development goals: Growth versus ecology on a finite planet. *Sustainable Development* 27 (5): 873–884.

Horton, Richard, Robert Beaglehole, Ruth Bonita, John Raeburn, Martin McKee, and Stig Wall. 2014. From public to planetary health: A manifesto. *Lancet* 383:847.

Hughes, David J., Rachel Alderdice, Christopher Cooney, Michael Kühl, Mathieu Pernice, Christian R. Voolstra, and David J. Suggett. 2020. Coral reef survival under accelerating ocean deoxygenation. *Nature Climate Change* 10:296–307.

International Association of Human-Animal Interaction Organizations (IAHAIO). 2018. *The IAHAIO Definitions for animal assisted intervention and guideline for wellness of animals involved in AAI*. IAHAIO White Paper 2014, updated 2018. Retrieved August 15 2022, https://iahaio.org/wp/wp-content/uploads/2021/01/iahaio-white-paper-2018-english.pdf

IUCN SSC. 2014. *IUCN SSC Guidelines for minimizing the negative impact to bats and other cave organisms from guano harvesting*. Ver. 1.0. IUCN, Gland.

Jenkins, Emily J., Audrey Simon, Nicholas Bachand, and Craig Stephen. 2015. Wildlife parasites in a one health world. *Trends in Parasitology* 31 (5): 174–180.

Keeling, Linda, Håkan Tunón, Gabriela Olmos Antillón, Charlotte Berg, Mike Jones, Leopoldo Stuardo, Janice Swanson, Anna Wallenbeck, Christoph Winckler, and Harry Blokhuis. 2019. Animal welfare and the united nations sustainable development goals. *Frontiers in Veterinary Science* 6:336.

Klimpel, Sven, and Heinz Mehlhorn. 2014. *Bats (Chiroptera) as vectors of diseases and parasites: Facts and myths*. New York: Springer.

Kunz, Thomas H., Elizabeth Braun, Dana de Torrez, Tatyana Lobova Bauer, and Theodore H. Fleming. 2011. Ecosystem services provided by bats. *Annals of the New York Academy of Sciences* 1223:1–38. https://doi.org/10.1111/j.1749-6632.2011.06004.x.

Kymlicka, Will. 2002. Contemporary political philosophy. 2nd ed. Oxford University Press: Oxford.

Lederman, Zohar. 2018. *Stamping out culling: The ethics of culling from a one health perspective*. Ph. D. thesis. Centre for Biomedical Ethics, National University of Singapore.

Lerner, Henrik. 2012. Virchow. In *Filosofi och medicin: Från Platon till Foucault*, ed. Lennart Nordenfelt, 135–150. Thales. [In Swedish].

Lerner, Henrik. 2013. Philosophical roots to the one medicine movement: An analysis of some relevant ideas by rudolf virchow and calvin schwabe with their modern implications. *Studia Philosophica Estonica* 6 (2): 97–109.

Lerner, Henrik. 2020. Health in non-human organisms. In *Explaining health across the sciences*, ed. J. Sholl and S. I. S. Rattan, 333–346. Cham, Switzerland: Springer.

Lerner, Henrik, and Charlotte Berg. 2015. The concept of health in one health and some practical implications for research and education: What is one health? *Infection Ecology & Epidemiology* 5:25300. https://doi.org/10.3402/iee.v5.25300.

Lerner, Henrik, and Jakob Zinsstag. 2021. Towards a healthy concept of health. In *One health: The theory and practice of integrated health approaches*, ed. J. Zinsstag, E. Schelling, L. Crump, M. Whittaker, M. Tanner, and C. Stephen, 2nd ed., 52–56. Wallingford, UK: CABI.

Lerner, Henrik, Bo Algers, Stefan Gunnarsson, and Anders Nordgren. 2011. Livestock's "short shadow"? Balancing mitigation of climate change against other values. In: Köfer, J., and Schobesberger, H. Animal hygiene and sustainable livestock production. Proceedings of the XVth international congress of the international society for animal hygiene, vienna, austria, 3–7 July 2011. Volume 1, pp. 213–214.

Levinson, Boris M. 1962. The dog as a "co-therapist". *Mental Hygiene* 46:59–65.

Lu, Manman, Xindong Wang, Huan Ye, Huimin Wang, Shan Qiu, Hongmao Zhang, Ying Liu, Jinhong Luo, and Jiang Feng. 2021. Does public fear that bats spread COVID-19 jeopardize bat conservation? *Biological Conservation* 254:108952.

Melaun, Christian, Antje Werblow, Markus Wilhelm Busch, Andrew Liston, and Sven Klimpel. 2014. Bats as potential reservoir hosts for vector-borne diseases. In *Bats (Chiroptera) as vectors of diseases and parasites: Facts and myths*, ed. Sven Klimpel and Heinz Mehlhorn, 25–61. New York: Springer.

Middleton, Jo, Jackie A. Cassell, Gavin Colthart, Francesca Dem, James Fairhead, Michael G. Head, Joao Inacio, Mavis Jimbudo, Moses Laman, Vojtech Novotny, Mika Peck, Jonah Philip, William Pomat, Shen Sui, Peter West-Oram, and Alan Stewart. 2020. Rationale, experience and ethical considerations underpinning integrated actions to further global goals for health and land biodiversity in papua new guinea. *Sustainability Science* 15:1653–1664.

Mills, Daniel, and Sophie Hall. 2014. Animal-assisted interventions: Making better use of the human-animal bond. *Veterinary Record* 174:269–273. https://doi.org/10.1136/vr.g1929.

Nagel, Thomas. 1974. What is it like to be a bat? *The Philosophical Review* 83:435–450.

Ng, Zenithson, Laura Morse, Julia Albright, Ann Viera, and Marcy Souza. 2018. Describing the use of animals in animal-assisted intervention research. *Journal of Applied Animal Welfare Science* 22 (4): 364–376.

Prata, Joana C., João P. da Costa, Isabel Lopes, Anthony L. Andrady, Armando C. Duarte, and Teresa Rocha-Santos. 2021. A one health perspective of the impacts of microplastics on animal, human and environmental health. *Science of the Total Environment* 777:146094.

Prescott, Susan, and Alan C. Logan. 2019. Planetary health: From the wellspring of holistic medicine to personal and public health imperative. *The Explorer* 15 (2): 98–106. https://doi.org/10.1016/j.explore.2018.09.002.

Queenan, Kevin, Julie Garnier, Liza Rosenbaum Nielsen, Sandra Buttigieg, Daniele de Meneghi, Martin Holmberg, Jakob Zinsstag, Simon Rüegg, Barbara Häsler, and Richard Kock. 2017. *Roadmap to a one health agenda 2030*. CAB Reviews 2017 12, No. 014.

Rockström, Johan, and Owen Gaffney. 2021. *Breaking boundaries: The science of our planet*. New York: DK Publishing, Penguin Random House.

Saint-Charles, Johanne, Jena Webb, Andres Sanchez, Hein Mallee, Berna van Wendel de Joode, and Hung Nguyen-Viet. 2014. Ecohealth as a field: Looking forward. *EcoHealth*. https://doi.org/10.1007/s10393-014-0930-2

Salwén, Håkan. 2003. *Hume's law: An essay on moral reasoning*. Ph. D. thesis. Stockholm Studies in Philosophy 25. Stockholm: Almqvist & Wiksell International.

Shafer Landau, Russ. 2012. The fundamentals of ethics. 2nd ed. New York and Oxford: Oxford University Press.

Sumaila, U. Rashid, and Travis C. Tai. 2020. End overfishing and increase the resilience of the ocean to climate change. *Frontiers in Marine Science* 7:523.

Szende, Jennifer. 2022. Relational value, land, and climate justice. *Journal of Global Ethics* 18 (1): 118–133.

Talukder, Byomkesh, Nilanjana Ganguli, Richard Matthew, Gary W. van Loon, Keith W. Hipel, and James Orbinski. 2022. Climate change-accelerated ocean biodiversity loss & associated planetary health impacts. *The Journal of Climate Change and Health* 6:100114.

Thomas, Keith. 1984. *Man and the Natural World: Changing Attitudes in England 1500–1800*. Penguin Books edition. London: Penguin Books.

Waltner-Toews, David. 2017. Zoonoses, One health and complexity: Wicked problems and constructive conflict. *Philosophical Transactions of the Royal Society B* 372:20160171.

Whitmee, Sarah, Andy Haines, Chris Beyrer, Frederick Boltz, Anthony G. Capon, Braulio Ferreira de Souza Dias, Alex Ezeh, Howard Frumkin, Peng Gong, Peter Head, Richard Horton, Georgina M. Mace, Robert Marten, Samuel S. Myers, Sania Nishtar, Steven A. Osofsky, Subhrendu K. Pattanayak, Montira J. Pongsiri, Cristina Romanelli, Agnes Soucat, Jeanette Vega, and Derek Yach. 2015. Safeguarding human health in the anthropocene epoch: Report of the rockefeller foundation-lancet commission on planetary health. *Lancet* 386:1973–2028.

Part I
Ethics and One Health Approaches

Chapter 2
The Values and Scope of One Health Approaches

> *Innovative and disruptive approaches are needed to address shared health threats such as climate change, urbanization, and pollution, because business as usual is insufficient to inspire the necessary actions to protect the health of one species without risking the health of another. (Stephen 2021, 13)*

Abstract This chapter, the first of the two empirical chapters, analyses the definitions, contributing sciences and core values of One Health approaches such as One Health, EcoHealth, Planetary Health, and One Welfare. These four have been chosen as the main ones among others. Both meta-ethical and ethical aspects found in core texts of the different approaches are summarized here. Meta-ethical aspects address issues such as views on moral status, how to include ethics in the approach, and what kind of values should be considered. Ethical aspects cover core concepts that a sufficient ethical framework for One Health approaches need to be considered. The grid in One Health approaches is broad. One has both to consider different levels (individual, population, and ecosystem) as well as species (at least humans, animals, and plants).

Keywords One Health · Planetary Health · EcoHealth · One Welfare · Definition · Values · Interdisciplinarity · Umbrella

---

A forerunner to this text was work done together with Charlotte Berg published in the journal *Frontiers in Veterinary Science* (Lerner and Berg 2017). The purpose of that work was to compare values and scientific scope of three main One Health approaches; One Health, EcoHealth, and Planetary Health. Although similar in structure and some of the content, the text has been rewritten, updated and extended solely by me to meet the aim of this book.

H. Lerner, *Ethics for One Health Approaches: A Roadmap for Future Directions*, Advancing Global Bioethics 22,
https://doi.org/10.1007/978-3-032-05239-1_2

2.1 Introduction

In this Chapter—the first of the two empirical chapters—I will demarcate the area by presenting the main different One Health approaches and their view on science, scientific fields, knowledge base, core values, and how other entities than humans are valued. Parts of this Chapter cover meta-ethical aspects such as views on moral status, how to include ethics in the approach, and what kind of values to consider, while other parts cover core concepts that a sufficient ethical framework for One Health approaches need to be considered.

2.1.1 *A Note on Method*

The material consists of core texts with emphasis on theoretical matters. These are books, scientific papers, and other literature which have had an impact on the approaches. I have extensively gathered these texts through literature searches, cited reference searches, reading reference lists, following several organizations mailing lists (where they highlight important works), attended conferences, as well as having discussions with several distinguished people in the field since approximately 2010. My choice of texts is therefore based on a fairly long period of interaction with this field where I have had time to evaluate and re-evaluate texts. Parts of my insights have already been presented elsewhere,[1] but this compilation is an updated comprehensive view of the field. Texts up to June 2023 are included. I am fully aware that crucial texts might have been missing, but I have tried to be as objective as possible. My own position in the field has been a broad view on One Health approaches acknowledging differences between the approaches, which also might influence my presentation here.

2.2 The Development of One Health Approaches

During the late twentieth century, an understanding of health as complex and interconnected between humans, animals and ecosystems evolved in several areas of science. As highlighted in the epigraph from Craig Stephen, analyzing multiple threats to human health revealed that a one-sided focus on humans is insufficient to solve these issues. The One Health approaches evolved.

In this chapter I will present each of the most influential approaches that deals with multispecies health and inter- or transdisciplinarity. They are fairly modern,

---

[1] See especially Lerner and Berg (2015, 2017) and Lerner (2012, 2013, 2019). I have since 2002 worked on the comparison of health definitions for humans, animals, and other organisms (Lerner 2008, 2020).

even if they have forerunners. The existing names were set in the twenty-first century. I will follow what seems to be the present consensus view, that there are three main versions of One Health approaches named One Health, EcoHealth and Planetary Health (Hill-Cawthorne 2019; Zinsstag et al. 2021; Mumford et al. 2023). One Welfare, which is sometimes said to be a part of One Health, is in my view something else than One Health and needs to stand on its own, because it has a focus on the welfare aspect of life rather than health. Although its focus on welfare instead of health, it still belongs to these approaches because health and welfare belong to a concept cluster[2] where the concepts are connected to each other. There are also other approaches such as Conservation Medicine, GeoHealth, One Biology and Zoobiquity. I mention some of them briefly in the footnotes to this chapter, in order to help the reader to understand their connection to the main approaches described in this chapter.[3]

I have treated the four approaches analyzed in this chapter as stand-alone approaches, which implies that each of them has their own particular set of principles, values and emphasis on different aspects, despite attempts to merge them into one. I will come back to this issue of merging in Sect. 2.7 where I compare the four approaches. There is no universal agreed definition for any of these approaches and therefore some versions of one approach might be merged successfully with another approach. Here, I want to develop the differences based on the key theoretical literature for the four approaches, to provide a comprehensive overview of the entire field.

To provide an empirical (descriptive) background for in-depth normative studies, I need to conduct two analyses. The first, presented in this chapter, will analyze the scope of the different approaches, core values and aspects within the approaches that influence how one would consider moral status. This chapter gives information that will be helpful in further meta-ethical and normative thinking for these approaches. The second, presented in Chap. 3, will analyze normative ethical attempts being made within the field of One Health approaches.

This chapter therefore deals more or less with the fundamental values necessary to consider in finding ethical frameworks for One Health approaches.

---

[2] This is evident in my thesis where I through analysis on concept clusters studied the concepts of health, well-being and welfare (Lerner 2008). Further concepts such as wellness and quality of life could also belong to such a cluster. In a cluster, concepts could be synonymous, partly overlapping, be defined by each other or be distinct from each other. This view differs from the family resemblance view proposed by Ludwig Wittgenstein and gives a more holistic view of a concept and its relations to other similar or familiar concepts. It's a way of mapping a concept.

[3] For an insightful paper that summarizes 18 scientific fields that considers at least some aspects of the connection between humans, ecosystems and health see Oestreicher et al. (2018).

2.3 One Health

Among the One Health approaches, the one with the oldest roots is One Health, stemming back to the nineteenth century and the theoretical works of One Medicine by Rudolf Virchow (1821–1902) combining medical and veterinary science (Lerner 2012, 2013).[4] The grandparents to the modern One Health are often regarded as Rudolf Virchow, William Osler (1849–1919), James Harlan Steele (1913–2013), and Calvin W. Schwabe (1927–2006).[5] Another forerunner might be Conservation Medicine.[6]

In the early twenty-first century the modern One Health approaches arise. In 2004, there was a conference to bring veterinary medicine, human medicine and ecology together. As a result of this gathering, a statement called the 12 Manhattan Principles was issued which linked humans, animals and ecosystems (Cook et al. 2004). This is regularly seen as the starting point for One Health and at least a small part has bearing on the ethical discussion and is further discussed in Sect. 3.3.1.1.

In 2005, British Medical Journal and Veterinary Record announced a joint publication series of papers starting in November that year to promote collaborations between human and veterinary medicine (Alder and Easton 2005).

2.3.1 *Definitions of the Approach*

Lerner and Berg (2017) distinguish a narrow and a wide One Health approach. The narrow approach is similar to the old One Medicine approach in contributing sciences.[7] A recent example of such a narrow definition of One Health is found in a handbook on the approach:

> One Health [can thus be defined] as any added value in terms of health of humans and animals, financial savings, social resilience and environmental sustainability achievable by the cooperation of human and veterinary medicine and other disciplines when compared to the two medicines and other disciplines working separately (Zinsstag et al. 2021, 16–17).

---

[4] Oestreicher et al. (2018) falsely places its origins in the 1950s.

[5] See for example Hill-Cawthorne (2019) and Kaplan et al. (2023) for such historical presentations.

[6] Conservation medicine might be an approach that could exist on its own. However, some of the original sources (Tabor 2002; Weinhold 2003) show similarities with what is now One Health but with more emphasis on ecological health and a narrower focus on animals only considering wildlife. Another source (Aguirre and Gómez 2009) rather resembles what is now Planetary Health. Aguirre (2009) holds a version of Conservation Medicine that shows similarities to EcoHealth, see Sect. 2.4 for further linkage between these two approaches.

[7] There are also attempts for introducing other terms for this approach, with a slight difference in focus, for example Zoobiquity (see Lerner and Berg (2015) for a discussion on this approach proposed by Natterson-Horowitz and Bowers (2013)).

Here, the narrowness relates to animal and human health, although interdisciplinary, the several other sciences proposed, such as economics and biological sciences, could indicate a wide approach (Lerner and Zinsstag 2021).

One example of a wider definition has recently been proposed by the One Health High Level Expert Panel of the joint coalition of WHO, FAO, WOAH and UNEP[8]:

> One Health is an integrated, unifying approach that aims to sustainably balance and optimize the health of people, animals and ecosystems.
>
> It recognizes the health of humans, domestic and wild animals, plants, and the wider environment (including ecosystems) are closely linked and inter-dependent.
>
> The approach mobilizes multiple sectors, disciplines and communities at varying levels of society to work together to foster well-being and tackle threats to health and ecosystems, while addressing the collective need for clean water, energy and air, safe and nutritious food, taking action on climate change, and contributing to sustainable development (WHO 2021).

Here, humans, animals, plants and ecosystems are more on par with a focus on a broad view of sustainable development. Also, crucial is the focus on multi-, inter- or transdisciplinarity and the value of connecting communities with societal sectors and research disciplines, which is often emphasized within the approach.

2.3.2 Contributing Sciences

The *narrow* version of One Health consists mainly of veterinarians and physicians (especially public health) and resembles its forerunner One Medicine (Lerner and Berg 2017). Proponents use the strong heritage from One Medicine and favor these two disciplines as the core disciplines in One Health. In this version, medicine, regardless of species, needs an inter- or transdisciplinary scientific bridge (Kaplan et al. 2023).

The *wide* version also includes ecology, agricultural sciences, other human health sciences including microbiology, health economics and sociology (Scoones and Forster 2009; Binot et al. 2015; Lerner and Berg 2017). The expansion into arts and humanities as well as the ecological field (including environmental health) is ongoing. The Manhattan Principles (Cook et al. 2004) in 2004 made the wide approach possible through the inclusion of ecology although recently ecology has become the forefront. In 2009, when swine flu was a pandemic, there was a call for a wider version of One Health including economists and social scientists (Scoones and Forster 2009). A recent bibliometric analysis (Miao et al. 2022) of research priorities within One Health showed that this widening has been the case. Although

[8] WHO is abbreviation for World Health Organization (United Nations), FAO for Food and Agriculture Organization of the United Nations, WOAH for World Organisation for Animal Health (formerly OIE, Office International des Epizooties), and UNEP for United Nations Environment Programme.

the natural sciences are still dominant (especially veterinary and human medicine), the researchers highlighted for example social sciences, economics, and nursing science as contributing to the approach.

The inclusion of ecological sciences together with the widening into arts and humanities makes One Health more like EcoHealth than before, at least when one analyzes contributing sciences.[9] This move has been a result of the critique (for example Natterson-Horowitz and Bowers (2013) and Roger et al. (2016)) of One Health in the narrow version.

The knowledge base is mainly western scientific in both the narrow and the wide approach (Lerner and Berg 2017), although recent developments of the approach include Indigenous peoples' and local communities' knowledge (Laing et al. 2023; Mumford et al. 2023).[10]

There are several proponents claiming that research must be either inter- or transdisciplinary rather than multidisciplinary in order to fall within the One Health approach (among the most recent suggestions, Rüegg et al. 2018; Lerner and Zinsstag 2021). Manlove et al. (2016) questioned the inter- or transdisciplinary claim by showing that several of the disciplines still worked in silos with very little collaboration outside their discipline. One solution to avoid the silos has been the proposal of a method to calculate what the authors call the OH-ness, a method evaluating how much inter- or transdisciplinarity that actually is involved (Rüegg et al. 2018). Research could be considered to be One Health when the level of inter- or transdiciplinarity is above a certain level according to their method of calculating the OH-ness. This is an example of the strong urge for inter- and transdisciplinary research within One Health.

2.3.3 Core Values

The change of term from One Medicine to One Health for the approach acknowledges a shift from a curative perspective towards a health perspective (Hill-Cawthorne 2019). Health is therefore a core value although one sometimes gets the impression that a lot of effort is on curing or avoiding disease. The narrow approach focuses on individual health and concerns only humans and animals. Here, there might be a close connection to human public health. The narrow version might be

[9]This has been disputed with the claim that although ecological sciences contribute, they are still not valued as the core sciences (Bruchhausen, 2019).

[10]To demarcate Indigenous peoples', First Nations', and local communities' knowledge several terms are used, for example Indigenous knowledge, local knowledge, traditional knowledge, and traditional ecological knowledge (Lerner and Tunón 2010).

One definition of traditional knowledge is by Fikret Berkes et al. (2000), "a cumulative body of knowledge, practice, and belief, evolving by adaptive processes and handed down through generations by cultural transmission, about the relationship of living beings (including humans) with one another and with their environment".

viewed just as an extension of human public health adding the value of animal health in those cases where both animals and humans can be influenced (Lerner and Berg 2017). If the narrow version treats humans and animals as equals, this version will be more true to the values of One Health.

The wide approach has besides individual health also focus on population health. In the wide approach ecosystems are added to groups of concern (Lerner and Berg 2017; Hill-Cawthorne 2019). Plants have often been neglected, and there has been calls for including plants (Destoumieux-Garzón et al. 2018). Nowadays plants (most probably also including fungi) are recognized as a group of concern, which is also reflected in the OHHLEP definition of One Health cited in Sect. 2.3.1.[11]

Moral status is ascribed to humans and animals in the narrow version. In the wide version moral status could be ascribed to humans, animals, plants, and ecosystems, although emphasis tends to be on humans and animals.

One phrase commonly used by scholars is that One Health deals with "the inextricable connection between humans, animals, and the environment".[12] Another way to put it is the emphasis on the "animal-human-ecosystem interface".[13] This is a value laden phrase, which implies that the relation between these three are important. They depend on each other, and this phrase opens for a similar valuation of all three. The term "inextricable connection" indicates that relations between these groups are important.

One early ethical value identified, as important in the expansion for a wider One Health approach, was equity (Scoones and Forster 2009). However, in that paper, equity was more of a buzz word without being fully analyzed. Although, there were no call for ethicists or philosophers in the paper, they asked for the inclusion of social scientists and economists (see further Sect. 3.2).

In the distinction of anthropocentrism, zoocentrism, biocentrism, or ecocentrism, both the narrow and wide versions of One Health tend to be zoocentric. Although the approach has been criticized for being anthropocentric, there is a strong emphasis on avoiding anthropocentrism. Although plants and ecosystems have the possibility to be ascribed values in the wide version, one can hardly say that One Health has turned into being biocentric or ecocentric either. Therefore, at the moment, one can see One Health as a zoocentric approach.

[11] One early inclusion of plants was proposed already in 1927 by Frederick Parkes Weber in British Medical Journal for the approach Comparative Medicine which is a forerunner to the One Health approach. His definition of comparative medicine was:

"'The study of disease and its treatment, by comparison in human beings, animals, and plants, and in relation to race or variety, sex, age, place in family, climate, season, soil, food, and mode of life- (as modified by occupation, habits, and surrounding influences).'

It must, of course, necessarily include the study of normal as well as pathological anatomy and physiology, and must partially depend on a variety of other scientific knowledge." (Parkes Weber 1927, 847).

[12] Used in several papers with slightly different phrasing but with similar meaning (see for example Bruchhausen 2019). This is also used in one of the definitions for EcoHealth, see Sect. 2.4.1.

[13] Used in several texts (for example Bruchhausen 2019).

2.4 EcoHealth

The origins of EcoHealth stem from the 1990s (Oestreicher et al. 2018) and already in 1996, the International Development Research Centre (IDRC) initiated an EcoHealth program (Lebel 2003). Ecohealth was developed and merged from the approaches of ecosystem health, global change and health, and Conservation Medicine (Charron 2012; Saint-Charles et al. 2014).[14] However, first in 2004, the Ecohealth journal established the field of EcoHealth (Saint-Charles et al. 2014). One of the most influential theorists in this approach is Jean Lebel (Hill-Cawthorne 2019). EcoHealth does not seem to have a narrow or wide version (Lerner and Berg 2017).

2.4.1 Definitions of the Approach

EcoHealth considers the health of humans, animals, and ecosystems. Environmental sustainability and socioeconomic stability are also included (Lerner and Berg 2017). Waltner-Toews defines that EcoHealth has as a goal.

> sustainable human and animal health and well-being, through healthier ecosystems (Waltner-Toews 2009, 519).

One can argue that there might be a narrow version indicated in Waltner-Toews' definition where emphasis seems to be more on humans and animals than on ecosystems. Animals and humans have both health and well-being as valuable concepts, while ecosystems only have health. However, health is attributed to all three and well-being could hardly be attributed to ecosystems. Therefore, I see this definition as similar to other definitions of EcoHealth in the field. In 2017, EcoHealth, the leading journal in the field, defined the approach as:

> EcoHealth is committed to fostering the health of humans, animals, and ecosystems and to conducting research which recognizes the inextricable linkages between the health of all species and their environments. A basic tenet held is that health and well-being cannot be sustained in a resource depleted, polluted, and socially unstable planet.[15]

A more recent characterization of EcoHealth is given by Hill-Cawthorne:

> Ecosystem health, or EcoHealth, is a systems-based approach to promoting health and well-being with a focus on social and ecological interactions. (Hill-Cawthorne 2019, 9).

---

[14] One example of a version of Conservation Medicine similar to EcoHealth is Aguirre (2009). See also footnote 6 for a further description on similarities between Conservation Medicine and One Health, EcoHealth and Planetary Health.

[15] https://ecohealth.net/en/ retrieved August 14, 2017.

The emphasis is on all living creatures, and the approach has also a more biological-ecological view than One Health. This makes it easier to argue that parasites, unicellular organisms, and potentionally viruses (whether they are defined as organisms or not)[16] have a value and should be protected (Lerner and Berg 2017).

2.4.2 Contributing Sciences

Early in the development of EcoHealth the demarcation of sciences included a broad range such as conservation and ecosystem management, rural and urban development and planning as well as veterinary medicine, human medicine, and public health practice (Wilcox et al. 2004). Ecologists have therefore been more influential in this approach than in, for example, One Health.

Besides western scientific knowledge, EcoHealth acknowledges Indigenous knowledge as an important source (Saint-Charles et al. 2014) and included it earlier than the other approaches. There has been more inclusion of social science and especially humanities within EcoHealth than the other approaches. Therefore, essays, poems, and more aesthetic visualizations are accepted scientific contributions (Lerner and Berg 2017).

Transdisciplinary research, together with participation, systems thinking, and a knowledge to action are leading perspectives within EcoHealth (Lebel 2003; Charron 2012). Participation is cooperation-based with the aim of consensus (Lebel 2003). This consensus includes scientists, policymakers and those conducting the fieldwork (Charron 2012; Saint-Charles et al. 2014).

2.4.3 Core Values

In the definition from the EcoHealth journal presented above (Sect. 2.4.1) health are ascribed to humans, animals, and ecosystems. For health, individual health has less importance, and the main emphasis is on the population level. In the characterization made by Hill-Cawthorne above this is phrased as "a focus on social and ecological interactions". Health could also be used as a metaphor ascribed to processes and ecosystems (Charron 2012). Besides health, well-being is also acknowledged as an important concept in all definitions.

Among other important values, sustainability is a core value and includes also ecological sustainability. When Jean Lebel (2003) analyses the concept of health in EcoHealth, he places health as depending on the environment, community, and economy. There is resemblance to the three-pillar model of sustainability now in use (see Adams 2006). Biodiversity is regarded as an important value in itself, which

[16] See Lerner (2020) for a brief analysis on the concept of organisms relating to One Health.

avoids sustainability concepts that are more anthropocentric such as the Brundtland concept of sustainability (Lerner and Berg 2017). This is a crucial difference between EcoHealth and Planetary Health, where the latter sees biodiversity as a means to human thriving (see Sect. 2.5.3; the sustainability concept will be further discussed in Sects. 4.2.1 and 6.2.3).

In the discussion on sustainable development, ecosystems, and health, an ethical value lifted is equity (Lebel 2003; Wilcox et al. 2004; Charron 2012; Zinsstag et al. 2012; Saint-Charles et al. 2014). Equity is specified as gender and social equity in the early definitions of the approach (Lebel 2003; Charron 2012).

Ecohealth, with its strong emphasis on ecology and population or ecosystem level, clearly indicates that this approach has a biocentric or ecocentric view.

2.5 Planetary Health

The roots of Planetary Health are placed in the 1970s, with theoretical elaboration by Trevor Hancock in the 1990s. One forerunner might be Conservation Medicine.[17] An early attempt named Planetary Health was put forward by Richard Horton et al. (2014). However, Planetary Health as an approach was established in 2015 by the Rockefeller Foundation-Lancet Commission on Planetary Health (Hill-Cawthorne 2019; Prescott and Logan 2019). Through this commission, Whitmee et al. (2015) introduced the approach in an extensive paper which is key to understanding it.[18]

2.5.1 Definitions of the Approach

There is a narrow and a wide version of Planetary Health (Lerner and Berg 2017). A narrow version is presented in the paper by Richard Horton et al. extending public health to planetary health. In this condensated qoute they elaborate on their view:

> Planetary health … emphasises people, not diseases, and equity, not the creation of unjust societies. We seek to minimise differences in health according to wealth, education, gender, and place. […] We need a new vision of cooperative and democratic action at all levels of society and a new principle of planetism and wellbeing for every person on this Earth—a

---

[17] See footnote 6.

[18] Another approach similar to Planetary Health is GeoHealth. The modern GeoHealth approach had forerunners in approaches called Earth Stewardship Initiative, Environmental Health Mission Area, and GeoHealth Initiative. In 2016, the establishment from the American Geophysical Union of a GeoHealth initiative together with a journal with the same name GeoHealth formed the approach (Almada et al. 2017). This approach consists of health sciences, ecology and earth sciences. Even space sciences contribute to the approach. Due to the emphasis on geosphere-biosphere interface (Dovie et al. 2022), a certain emphasis on abiotic aspects is evident in GeoHealth. Dovie et al. (2022) places GeoHealth within the One Health approach, while Almada et al. (2017) suggests that Planetary Health and GeoHealth could be merged.

> principle that asserts that we must conserve, sustain, and make resilient the planetary and human systems on which health depends by giving priority to the wellbeing of all. (Horton et al. 2014, 847).

The central wide definition of this approach is the one proposed by the Rockefeller Foundation-Lancet Commission on planetary health:

> Our definition of planetary health is the achievement of the highest attainable standard of health, well-being, and equity worldwide through judicious attention to the human systems—political, economic, and social—that shape the future of humanity and the Earth's natural systems that define the safe environmental limits within which humanity can flourish. (Whitmee et al. 2015, 1978).

Here a strong link between humanity and earth's natural systems is established. An optimal level of health, well-being and equity is aimed for, and the environment should be protected at a level that is needed to safeguard humanity.

2.5.2 Contributing Sciences

In the early narrow version (Horton et al. 2014), Planetary Health seemed to be very much an extension of global public health because of the suggested contributing sciences that were only health professionals and scientists. In this suggestion neither veterinary medicine nor ecology were explicitly mentioned (Lerner and Berg 2017). A year later, the influential report from Whitmee et al. (2015) defined the approach with a more interdisciplinary collaboration with contributions from human medicine, ecology and environmental sciences, agricultural sciences (including plant and animal production sciences), marine sciences, energy sciences and economy (Lerner and Berg 2017). There is a lack of acknowledgement of veterinary medicine as a core discipline within the approach although zoonotic and vector-borne diseases are mentioned as threats to humans. One reason might be that this approach has had more emphasis on climate change, sustainabilty and health (Whitmee et al. 2015; de Paula 2021) rather than zoonoses and health.

The knowledge base is western scientific for both the narrow and the wide approach (Lerner and Berg 2017), although Nicole Redvers et al. (2020, 2022) has developed the approach to also include Indigenous peoples' knowledge.

Seen as a transdisciplinary field, there still have been calls for more transdisciplinarity within the approach (Hill-Cawthorne 2019).

2.5.3 *Core Values*

In both the narrow and wide versions focus are on human health, which is recognized at both the individual and population level. What could be loose definitions of health for ecosystems, or the planet, are a view on earth's natural systems that protects humans (Whitmee et al. 2015).

For both the narrow and wide version, sustainability is one of the important values, but this concept is based on the Brundtland version implying sustainability for living and future human generations (compared to EcoHealth's interpretation of the concept). Biodiversity and natural resources are therefore valued as prerequisites for humans. An ethical concept, equity in health, has been a goal within this approach since its establishment. This is understood as equity among humans despite differences in socioeconomic, regional or gender factors (Lerner and Berg 2017).

Only humans seem to have moral status within this approach, although biodiversity is valued. This seems to be to safeguard human existence and therefore being an instrumental value for humans.

I have earlier (Lerner and Berg 2017) questioned if this approach is a One Health approach due to its lack of an interspecies focus in values. It could rather be characterized as a wider version of public or global health. This is obvious for both of the early definitions (Horton et al. 2014; Whitmee et al. 2015) mentioned in Sect. 2.5.1, but holds in my view even for more recent developments on the approach (such as Myers et al. 2021; de Paula 2021) and Hill-Cawthorne (2019) sees it as a successor to both public and global health just adding environmental drivers. Altogether, Planetary Health in both its narrow and wide sense is anthropocentric. However, as mentioned at the start of the Chapter, the consensus view is that it belongs to the One Health approaches mainly due to its interdisciplinarity across the medical and ecology fields. Therefore, it is also a part of this analysis.

2.6 One Welfare

One of the earliest to coin One Welfare as a term was Daniel Mills (Anonymous 2012; Mills and Hall 2014).[19] One Welfare developed from the One Health approach and focused rather on the concept of welfare than the concept of health. In an early attempt, Colonius and Earley (2013) proposed that animal welfare, human welfare,

[19] It is here on its place to mention the term "Zooeyia" which was coined by Kate Hodgson and Marcia Darling, originally suggested to be included in One Health (Hodgson and Darling 2011) and later in One Welfare (Williamson et al. 2022). The term was developed in a narrow version of One Health and defined as "the positive benefits to human health from interacting with animal, focusing on the companion animal" and have recently gained further interest among some scholars within AAT studies. However, Zooeyia has been criticized for its anthropocentric focus on human benefits (Van Patter et al. 2023) which makes the term in conflict with core values of inter- or

and social welfare should no longer be seen as separate disciplines. The main theory of the approach has been developed by Rebecca García Pinillos (García Pinillos et al. 2016; García Pinillos 2018).[20]

2.6.1 *Definitions of the Approach*

In García Pinillos (2018) thorough framework of One Welfare, she defines the approach in relation to One Health as focusing on the animal welfare and human well-being aspect rather than the health aspect. Initially, the connection between One Welfare and One Health was somewhat open for discussion, as she states:

> While some are of the view that One Welfare is truly under the umbrella of One Health, others believe that One Welfare is broader and encompasses One Health (García Pinillos 2018, 10).

Her own view seems to be that they are complimentary, where One Health focuses on health aspects and One Welfare on welfare or well-being aspects (García Pinillos 2021). Still, a lot of values as well as demarcation of sciences will be similar between One Health and One Welfare. García Pinillos wants to avoid a single-sentence definition of the approach but the closest to such a definition is the following:

> One Welfare describes the interrelationships between animal welfare, human well-being and the physical and social environment (García Pinillos 2018, 12).

Colonius and Earley (2013) suggest in a similar way that the approach focuses on social, human, and animal welfare. And in a talk in Guadalajara in 2016, inspired by Colonius and Earley and the paper by García Pinillos et al. (2016), David Fraser proposed that the approach is based on

> … interconnections between human welfare, animal welfare and the integrity of the environment[21]

In the manuscript to the talk he interpreted welfare as flourishing:

---

transdisciplinarity in both One Health and One Welfare. Therefore, the framework of Zooeyia, for example described in Williamson et al. (2022), will not be considered in this book.

[20] A further development in terms has been the "One Biology" suggestion by Tarazona et al. (2019). One biology is more of a justification of why One Health and One Welfare are proper approaches and not another slightly different approach. Only animals and humans are included within One Biology, which implies that only the narrow version of One Health is considered under this umbrella concept.

[21] Abstract by David Fraser, 2016, What do we mean by "one welfare"? Guadalajara Fourth Global Conference on Animal Welfare, World Organisation for Animal Health, Dec 6–8, 2016, Guadalajara, Mexico. Personal communication from the author.

> Hence, one welfare means that people, animals and the ecological systems of the world are flourishing, and that the flourishing of each component is related to the flourishing of the others.[22]

2.6.2 *Contributing Sciences*

For One Welfare there have been different suggestions for the contributing sciences. Colonius and Earley (2013) suggest that evolutionary geneticists, behavioural ecologists, environmental scientists, neuroscientists, economists, social scientists, and public administration professionals could be valuable within this approach. David Fraser stated that veterinary and animal protection services, human health, and social services should be included.[23]

The knowledge base is western scientific. Multi- and interdisciplinarity is favored (García Pinillos 2018). García Pinillos three main contributing science areas are animal welfare, human well-being, and environmental conservation research, without further specification of scientific disciplines.

2.6.3 *Core Values*

In One Welfare, focus is on humans and animals rather than ecosystems (Colonius and Earley 2013; Mills and Hall 2014; García Pinillos 2018), therefore moral status seems to be ascribed only to animals and humans. In the first mention of the approach (Anonymous 2012) the suggestion was to expand the focus on physical health towards including also mental health for both animals and humans. Obviously, mental and social aspects of health are the focus within the health concept of this approach. The core values differs somewhat among the proponents. Colonius and Earley (2013) has animal welfare, human welfare and social welfare. Mills and Hall (2014) has the welfare of animals and humans. García Pinillos (2018, 13) defines One Welfare as having three core values, animal welfare, human well-being and environment conservation.

Ethics has been emphasized in this approach. Colonius and Earley (2013) list it among the issues to be solved, although not including philosophers or ethicists in the list of contributing disciplines. Among ethical issues that are pinpointed within the approach is for example minimizing human and animal abuse and neglect, finding animal assisted interventions that are ethically sound for both humans and animals, and providing animal welfare also for wild animals when sustainability is opt

---

[22] Written in the manuscript for the lecture held by David Fraser at the 2016 conference in Guadalajara, Mexico arranged by the World Organisation for Animal Health. Personal communication from the author.

[23] In the manuscript to the lecture held by David Fraser at the 2016 conference in Guadalajara, Mexico.

for (García Pinillos 2018). Avoidance of suffering is therefore core, but also other aspects of harm. Sustainability is also central and understood as the three-pillar model. These aspects will be further developed in Chap. 3 where I discuss ethical attempts already being made within the approaches.

Although García Pinillos include conservation of the environment, One Welfare seem to consider moral status in animals and humans. With its core focus on well-being and welfare in animals and humans this approach is zoocentric.

2.7 Comparison of the One Health Approaches

There are several ways to visualize similarities and differences between the One Health approaches. Here I will mention two ways. The first is the so-called umbrellas, which are umbrella depictions which show a comprehensive view of an approach. I will describe that method in Sect. 2.7.1. Another way is to use a comparative table which I have compiled and presented in Table 2.1. The table is based on Sects. 2.3, 2.4, 2.5, 2.6.

2.7.1 The Umbrellas

One way to get impact and visualize One Health approaches has been by using an umbrella depiction. The umbrella for One Health were developed by One Health Sweden in collaboration with One Health Initiative Autonomous *pro bono* team and designed by Charlotte Berg (Lerner and Berg 2015).[24] The umbrella showed disciplines of science within the approach, core aspects of the concept of health as well as specific research issues. This umbrella depicts a wide One Health approach because it has human, animal and ecosystem health as core aspects.

The umbrella for One Welfare was developed within the One Welfare approach and appeared first in 2016 designed by R. Held and then again in 2018 (García Pinillos et al. 2016; García Pinillos 2018, 13). This umbrella showed core aspects of the concept of welfare, core research areas and specific research issues. Core aspects of One Welfare are the connection between animal welfare, human well-being and environment conservation. *Noteworthy is that neither of these umbrellas include ethics or philosophy.*

One crucial aspect is that these umbrellas have been developed within the approaches and not from outside the approach. EcoHealth and Planetary Health have no umbrella depictions. They are possible to do, but if they should be similar to the other two existing, they need to be developed within their approaches.

[24] There are other umbrellas present, but they are less structured and lack the distinctive difference between different aspects of the approaches that I here develop for the One Health and One Welfare umbrella.

Table 2.1 Comparison of the four main approaches. See text for details[a]

	One Health		EcoHealth	Planetary Health		One Welfare
	Narrow	Wide		Narrow	Wide	
Human contributing sciences	Public health	Public health Human medicine Molecular and microbiology Health economics Social sciences	Public health Human medicine Rural and urban development and planning Social sciences Anthropology	Public health Human medicine	Human medicine Economy Energy Natural resources	Human health Human well-being Neuroscientists Economy Social sciences Public administration professionals
Animal contributing sciences	Veterinary medicine	Veterinary medicine	Veterinary medicine	–	Agricultural sciences (including plant and animal production sciences)	Veterinary and animal protection services Animal welfare
Ecosystem contributing sciences	–	Environmental health Ecology	Conservation and ecosystem management	–	Ecology Other environmental sciences (including climate and biodiversity research) Marine sciences	Behavioural ecologists Environmental scientists Environmental conservation research Evolutionary geneticists
Knowledge base	Western scientific	Western scientific Indigenous knowledge	Western scientific Indigenous knowledge	Western scientific	Western scientific Indigenous knowledge	Western scientific
Disciplinarity	Inter- or transdisciplinarity	Inter- or transdisciplinarity	Transdisciplinarity	Transdisciplinarity	Transdisciplinarity	Multi- or interdisciplinarity

	One Health		EcoHealth	Planetary Health		One Welfare
	Narrow	Wide		Narrow	Wide	
View of health	Individual	Individual Population	Population	Individual Population	Individual Population	Individual
Moral status	Humans Animals	Humans Animals Plants Ecosystems	Humans Animals Ecosystems	Humans	Humans	Humans Animals
Other important values	Equity	Equity	Equity Biodiversity Sustainability (for humans, animals, ecosystems)	Equity Sustainability (for humans)	Equity Sustainability (for humans)	Avoid suffering Welfare Sustainability
Perspective	Zoocentrism	Zoocentrism	Ecocentrism	Anthropocentrism	Anthropocentrism	Zoocentrism

[a]An earlier version of this table also appears in Lerner and Berg (2017). This table is revised, updated and extended

The strength of the use of an umbrella depiction of the approach is that the umbrella shows what is included within the approach. One reads the umbrella top-down. Closest to the top are the more fundamentally scientific aspects such as core scientific disciplines or research areas, and aspects of core concepts. These show the width of the umbrella, the scope of interdisciplinarity and core values as the width of health. Below these, there are more tentative examples shown of possible research issues. These could be crucial at the moment or possible pathways for the future.

In the future, if all approaches have these umbrella depictions, they could be used to easily explain each approach and serve as a basis for comparison between the approaches. One could then have these depictions as help when one wants to choose a scientific approach.

2.7.2 *Table for Comparison*

There have been several attempts to compare different approaches (for example Lerner and Berg 2017; Oestreicher et al. 2018). Table 2.1 comprehends what already has been stated in this chapter for the four main One Health approaches regarding core contributing sciences (divided on research on humans, animals, and ecosystems), knowledge base, disciplinarity, view of health, other important values and main ethical perspective (anthropocentrism, zoocentrism, biocentrism, or ecocentrism). Those approaches containing a narrow and a wide approach are separated in this table.

2.7.3 *Merging the Approaches*

Although each approach has somewhat different origin, values and scientific disciplines, there are still many similarities. Therefore, there have been suggestions for a merge of these approaches. One Health and EcoHealth have been suggested to be merged (Zinsstag et al. 2012, Roger et al. 2016) and this was also a main discussion at the One Health EcoHealth conference in Melbourne, Australia in December 2016. Others, like Hill-Cawthorne (2019) suggest that planetary health is overarching One Health and EcoHealth. This is contested if one agrees with the claim made originally by Lerner and Berg (2017) that Planetary Health is different, because it is anthropocentric (see Sect. 2.5). A caution mark is necessary. Although we call them One Health approaches, there might be a misconception that they all should aim for one similar approach, one definition of health or one ethical theory (Hinchliffe 2015; Lerner and Berg 2017). The reason behind the differences could be tracked down in fundamental differences in values and beliefs, which might be lost when merged. To proper solve this, one must adopt a view of plurality, so that these approaches could be working together towards the same goal.

2.7.4 *Species or Level Differences*

Where would ethics apply and how wide is the scope for the One Health approaches? For organisms, when already humans, animals and plants are considered, a reasonable wide suggestion is that all kingdoms of organisms should be included and maybe distinguished in ethical discussions. Viruses are according to some theorists not considered organisms and might be left out (see Lerner 2020). Inanimate objects, such as stone and rocks, are left out from this view since they have no life or reproduction.

For moral status one often chooses an aspect such as ability to suffer, rationality or life as a demarcation for those included. In One Health approaches one could work with moral status either in a top-down approach choosing either suffering, rationality, or something else that is primarily human but shared by other species. Another way is to use a bottom-up approach starting, for example with life, which could be shared by all.

Values could be attached to different levels.[25] One value that has been considered specifically in One Health approaches is health and could be a help to where values could be attributed. Health is attributed to individuals, populations and ecosystems (Lerner and Berg 2015, 2017). Organ or genes, lower levels than individuals, are not considered as entities to define health for. The species level is also important in the more ecological versions of the approaches. Species is a troublesome level, because it can be interpreted as both a systemic level (such as ecosystem) or a congregation of individuals (such as a congregation of populations that each consists of individuals). In the discussion of how to define health there is an urge to find a concept that could be attributed to all groups and levels of One Health (Stephen 2014; Döring et al. 2015; Lerner 2017, 2019, 2020). If this is the case for this core value of all approaches, I consider it relevant to also assume that one can deal with ethics like this, trying to find an ethical system that is comprehensive for all levels and organisms.[26]

2.8 Summary

In this Chapter I have demarcated the main different approaches and their view on science, scientific fields, knowledge base, disciplinarity, core values, and moral status. Moral status seems to differ between the approaches, but in the widest version

[25] Which is sometimes called scales. I will use levels throughout the book.

[26] There is a strand of critique of widening One Health approaches so that they become too broad, losing their specificity and become a theory of everything which is so diluted, so it loses its applicability. For example, one might end up in stating that everything is linked to health and therefore important to consider. These concerns have been raised by both scholars favoring and questioning One Health approaches (see for example Lerner and Berg 2017; Morar and Skorburg 2018; Bruchhausen 2019).

humans, animals, plants, and ecosystems could be seen as valuable. Health, which is one of the core values could be attributed to individual, population, and ecosystem level. Equity is one ethical concept that has been proposed in One Health, EcoHealth, and Planetary Health as important to analyze further. One Welfare aims for minimizing suffering and avoiding abuse and neglect in animals and humans.

The phrase "the inextricable connection", which is present in One Health and in one of the definitions of EcoHealth, might indicate that a relational ethic should be opt for. For the three health approaches—One Health, EcoHealth, and Planetary Health—recent development in including knowledge from Indigenous peoples and local communities makes them more uniform and facilitates a possible future merge. This might also imply that one needs to find a pluralistic ethic that acknowledges different value systems.

One can argue that differences between the approaches might hinder a future merge into one approach. Rather, one finds that the approaches might belong to fundamentally different strands of ethics, at least in some of their versions. One can argue that Planetary Health is primarily an anthropocentric approach, One Health and One Welfare zoocentric, and EcoHealth ecocentric.

In the remainder of the book, I must consider anthropocentric, zoocentric, biocentric and ecocentric ethical frameworks as well as values based on a broad cultural and religious understanding including both western scientific' values and Indigenous peoples' values. In the next chapter, I will turn to the proposals of ethics being made in the field of One Health approaches.

References

Adams, W.M. 2006. *The future of sustainability: Re-thinking environment and development in the twenty-first century*. Report of the IUCN Renowned Thinkers Meeting, 29–31 January 2006. https://portals.iucn.org/library/sites/library/files/documents/Rep-2006-002.pdf

Aguirre, A. Alonso. 2009. Wild canids as sentinels of ecological health: A conservation medicine perspective. *Parasites & Vectors* 2 (Suppl 1): S7.

Aguirre, A. Alonso, and A. Gómez. 2009. Essential veterinary education in conservation medicine and ecosystem health: A global perspective. *Revue Scientifique et Technique (International Office of Epizootics)* 28 (2): 597–603.

Alder, Martin, and Graham Easton. 2005. Human and veterinary medicine. *British Medical Journal* 330:858–859.

Almada, Amalia A., Christopher D. Golden, Steven A. Osofsky, and Samuel S. Myers. 2017. A case for Planetary Health/Geohealth. *GeoHealth* 1:75–78.

Anonymous. 2012. Thinking beyond one health. *Veterinary Record*. https://doi.org/10.1136/vr.e3116.

Berkes, Fikret, Johan Colding, and Carl Folke. 2000. Rediscovery of traditional ecological knowledge as adaptive management. *Ecological Applications* 10 (5): 1251–1262.

Binot, Aurelie, Raphaël Duboz, Panomsak Promburom, Waraphon Phimpraphai, Julien Cappelle, Claire Lajaunie, Flavie Luce Goutard, Tanu Pinyopummintr, Muriel Figuié, and François Louis Roger. 2015. A framework to promote collective action within the One Health community of practice: Using participatory modelling to enable interdisciplinary, cross-sectoral and multi-level integration. *One Health* 1:44–48.

Bruchhausen, Walter. 2019. Emerging global health approaches at the human-animal interface: Conceptual and historical issues of one health. In *Emerging global health approaches at the human-animal interface*, 1–32. IGI Global.

Charron, Dominique F. 2012. Ecohealth: Origins and approach. In: *Ecohealth research in practice: Innovative applications of an ecosystem approach to health,* ed. Dominique F. Charron, 1–30. Insight and innovation in international development 1. Ottawa: International Development Research Centre.

Colonius, Tristan J., and Rosemary W. Earley. 2013. One welfare: A call to develop a broader framework of thought and action. *JAVMA* 242 (3): 309–310.

Cook, Robert A., William B. Karesh, and Steven A. Osofsky. 2004. *The manhattan principles on "one world, one health"*. http://www.wcs-ahead.org/manhattan_principles.html

de Paula, Nicole. 2021. *Breaking the silos for planetary health: A roadmap for a resilient post-pandemic world.* Singapore: Palgrave Macmillan.

Destoumieux-Garzón, Delphine, Patrick Mavingui, Gilles Boetsch, Jérome Boissier, Frédéric Darriet, Priscilla Duboz, Clémentine Fritsch, Patrick Giradoux, Frédérique Le Roux, Serge Morand, Christine Paillard, Dominique Pontier, Cédric Sueur, and Yann Voituron. 2018. The One Health concept: 10 years old and a long road ahead. *Frontiers in Veterinary Science* 5:14.

Döring, Thomas F., Anja Viewiger, Marco Pautasso, Mette Vaarst, Maria R. Finckh, and Martin S. Wolfe. 2015. Resilience as a universal criterion of health. *Journal of the Science of Food and Agriculture* 95:455–465.

Dovie, Delali B. K., Michael K. Miyittah, Daniel E. Dodor, Mawuli Dzodzomenyo, Aaron K. Christian, Reuben Tete-Larbi, Samuel N. A. Codjoe, and Ayaga A. Bawah. 2022. Earth system's gatekeeping of "one health" approach to manage climate-sensitive infectious diseases. *GeoHealth* 6:e2021GH000543.

García Pinillos, Rebeca. 2018. *One welfare: A framework to improve animal welfare and human well-being*. Wallingford: CABI.

García Pinillos, Rebeca. 2021. One welfare impacts of COVID-19: A summary of key highlights within the one welfare framework. *Applied Animal Behaviour Science* 236:105262.

García Pinillos, Rebeca, M. Appleby, X. Manteca, F. Scott-Park, C. Smith, and A. Velarde. 2016. One Welfare: A platform for improving human and animal welfare. *Veterinary Record*. https://doi.org/10.1136/vr.i5470.

Hill-Cawthorne, Grant A. 2019. One Health/EcoHealth/Planetary Health and their evolution. In *One planet, one health*, ed. Merrilyn Walton, 1–19. Sydney: Sydney University Press.

Hinchliffe, Steve. 2015. More than one world, more than one health: Re-configuring interspecies health. *Social Science & Medicine* 129:28–35.

Hodgson, Kate, and Marcia Darling. 2011. Zooeyia: An essential component of "One Health". *Canadian Veterinary Journal* 52:189–191.

Horton, Richard, Robert Beaglehole, Ruth Bonita, John Raeburn, Martin McKee, and Stig Wall. 2014. From public to planetary health: A manifesto. *Lancet* 383:847.

Kaplan, Bruce, Laura H. Kahn, Thomas P. Monath, Lisa A. Conti, Thomas M. Yuill, Helena J. Chapman, Craig N. Carter, Becky Barrentine, Richard Seifman, and Jack P. Wodall. 2023. 'One Medicine–One Health': An historic perspective. *World Medical Journal* 69 (2): 18–27.

Laing, Gabrielle, Eleanor Duffy, Neil Anderson, Nicolas Antoine-Moussiaux, Maurizio Aragrande, Caetano Luiz Beber, John Berezowski, Elena Boriani, Massimo Canali, Luis Pedro Carmo, Ilias Chantziaras, Glen Cousquer, Daniele De Meneghi, Ana Gloria Rodrigues Sanches da Fonseca, Julie Garnier, Martin Hitziger, Thomas Jaenisch, Hans Keune, Claire Lajaunie, Lorena Franco Martinez, Rebecca Maudling, Marie K. McIntyre, Barry J. McMahon, Alberto Munoz Prieto, Liza Rosenbaum Nielsen, Ranya Özçelik, John W. A. Rossen, Simon R. Rüegg, Sara Savić, Margarida Pires Simoes, Deborah J. Thomson, Laura Tomassone, Asta Tvarijonaviciute, Manuela Vilhena, Barbara Vogler, and Barbara Häsler. 2023. Advancing one health: Updated core competencies. *CABI One Health*. https://doi.org/10.1079/cabionehealth.2023.0002.

Lebel, Jean. 2003. *Health: An ecosystem approach*. Focus. Ottawa: International Development Research Centre. Available from: https://www.idrc.ca/en/book/infocus-health-ecosystem-approach

Lerner, Henrik. 2008. *The concepts of health, well-being and welfare as applied to animals. A Philosophical analysis of the concepts with regard to the differences between animals*. Ph. D. Thesis. Linköping: Department of Medical and Health Sciences, Division of Health and Society, Linköping University. Linköping Studies in Arts and Science No. 438.

Lerner, Henrik. 2012. Virchow. In *Filosofi och medicin: från Platon till Foucault*, ed. Lennart Nordenfelt, 135–150. Stockholm: Thales. [In Swedish].

Lerner, Henrik. 2013. The philosophical roots of the "One Medicine" movement: An analysis of some relevant ideas by Rudolf Virchow and Calvin Schwabe with their modern implications. *Studia Philosophica Estonica* 6 (2): 97–109.

Lerner, Henrik. 2017. Conceptions of health and disease in plants and animals. In *Handbook of the philosophy of medicine*, ed. T. Schramme and S. Edwards, vol. 1, 287–301. Dordrecht, the Netherlands: Springer.

Lerner, Henrik. 2019. A critical analysis of definitions of health as balance in a One Health perspective. *Medicine, Health Care and Philosophy* 22 (3): 453–461.

Lerner, Henrik. 2020. Health in non-human organisms. In *Explaining health across the sciences*, ed. J. Sholl and S. I. S. Rattan, 333–346. Cham, Switzerland: Springer.

Lerner, Henrik, and Charlotte Berg. 2015. The concept of health in One Health and some practical implications for research and education: What is One Health? *Infection Ecology & Epidemiology* 5:25300. https://doi.org/10.3402/iee.v5.25300.

Lerner, Henrik, and Charlotte Berg. 2017. A comparison of three holistic approaches to health: One Health, EcoHealth, and Planetary Health. *Frontiers in Veterinary Science* 4:163. https://doi.org/10.3389/fvets.2017.00163.

Lerner, Henrik, and Håkan Tunón. 2010. Vad är traditionell och lokal kunskap? In *Nycklar till kunskap. Om människans bruk av naturen,* ed. Håkan Tunón, and Anna Dahlström, 41–57. Stockholm: Centrum för biologisk mångfald, Uppsala and Kungl. Skogs och Lantbruksakademien. [In Swedish].

Lerner, Henrik, and Jakob Zinsstag. 2021. Towards a healthy concept of health. In One Health: The theory and practice of integrated health approaches, ed. Zinsstag, J, E. Schelling, L. Crump, M. Whittaker, M. Tanner, and C. Stephen, 52–56. 2nd ed. Wallingford: CABI.

Manlove, Kezia R., Josephine G. Walker, Meggan E. Craft, Kathryn P. Huyvaert, Maxwell B. Joseph, Ryan S. Miller, Pauline Nol, Kelly A. Patyk, Daniel O'Brien, Daniel P. Walsh, and Paul C. Cross. 2016. "One Health" or three? Publication silos among the one health disciplines. *PLoS Biology* 14 (4): e1002448.

Miao, Liyuan, Hao Li, Wei Ding, Shenning Lu, Shuning Pan, Xiaokui Guo, XiaoNong Zhou, and Duoquan Wang. 2022. Research priorities on one health: A bibliometric analysis. *Frontiers in Public Health* 10:889854.

Mills, Daniel, and Sophie Hall. 2014. Animal-assisted interventions: Making better use of the human-animal bond. *Veterinary Record* 174:269–273.

Morar, Nicolae, and Joshua A. Skorburg. 2018. Bioethics and the hypothesis of extended health. *Kennedy Institute of Ethics Journal* 28 (3): 341–376.

Mumford, Elizabeth L., Deniss J. Martinez, Karli Tyance-Hassell, Alasdair Cook, Gail R. Hansen, Ronald Labonté, Jonna A. K. Mazet, Elaine C. Mumford, David M. Rizzo, Eri Togami, Arioene Vreedzaam, and John Parish-Sprowl. 2023. Evolution and expansion of the One Health approach to promote sustainable and resilient health and well-being: A call to action. *Frontiers in Public Health* 10:1056459.

Myers, Samuel S., Jeremy I. Pivor, and Antonio M. Saraiva. 2021. The São Paolo declaration on planetary health. *Lancet* 398:1299. including Supplementary appendix.

Natterson-Horowitz, B., and K. Bowers. 2013. *Zoobiquity: The astonishing connection between human and animal health*. New York: Vintage Books.

Oestreicher, Jordan S., Chris Buse, Ben Brisbois, Rebecca Patrick, Aaron Jenkins, Jonathan Kingsley, Renata Távora, and Lendra Fatorelli. 2018. Where ecosystems, people and health meet: Academic traditions and emerging fields for research and practice. *Sustentabilidade em Debate – Brasilia* 9 (1): 45–65.

Parkes Weber, Frederick. 1927. What is comparative medicine? *British Medical Journal* 2 (3487): 847.

Prescott, Susan, and Alan C. Logan. 2019. Planetary Health: From the wellspring of holistic medicine to personal and public health imperative. *The Explorer* 15 (2): 98–106. https://doi.org/10.1016/j.explore.2018.09.002.

Redvers, Nicole, Anne Poelina, Clinton Schultz, Daniel M. Kobei, Cicilia Githaiga, Marlikka Perdrisat, Donald Prince, and Be'sha Blondin. 2020. Indigenous natural and first law in Planetary Health. *Challenges* 11:29.

Redvers, Nicole, Yuria Celidwen, Clinton Schultz, Ojistoh Horn, Cicilia Githaiga, Melissa Vera, Marlikka Perdrisat, Lynn Mad Plume, Daniel Kobei, Myrna Cunningham Kain, Anne Poelina, Juan Nelson Rojas, and Be'sha Blondin. 2022. The determinants of planetary health: An Indigenous consensus perspective. *Lancet Planetary Health* 6:e156–e163.

Roger, François, Alexandre Caron, Serge Morand, Miguel Pedrono, Michel de Garine-Wichatitsky, Veronique Chevalier, Annelise Tran, Nicolas Gaidet, Muriel Figuié, Marie-Noël de Visscher, and Aurélie Binot. 2016. One Health and EcoHealth: The same wine in different bottles? *Infection Ecology and Epidemiology* 6:30978.

Rüegg, Simon R., Liza Rosenbaum Nielsen, Sandra C. Buttigieg, Mijalche Santa, Maurizio Aragrande, Massimo Canali, Timothy Ehlinger, Ilias Chantziaras, Elena Boriani, Miroslav Radeski, Mieghan Bruce, Kevin Queenan, and Barbara Häsler. 2018. A systems approach to evaluate one health initiatives. *Frontiers in Veterinary Science* 5:23.

Saint-Charles, Johanne, Jena Webb, Andres Sanchez, Hein Mallee, Berna van Wendel de Joode, and Hung Nguyen-Viet. 2014. Ecohealth as a field: Looking forward. *EcoHealth* 11:300–307.

Scoones, Ian, and Paul Forster. 2009. One World, One Health? *Rural* 21 (6): 22–24.

Stephen, Craig. 2014. Toward a modernized definition of wildlife health. *Journal of Wildlife Diseases* 50 (3): 427–430.

Stephen, Craig. 2021. The call to action. In *Animals, health, and society: Health promotion, harm reduction, and health equity in a one health world*, ed. Craig Stephen, 3–16. Boca Raton, London, and New York: CRC Press.

Tabor, Gary M. 2002. Defining conservation medicine. In *Conservation medicine: Ecological health in practice*, ed. A. Alonso Aguirre, Richard S. Ostfeld, Gary M. Tabor, Carol House, and Mary C. Pearl, 8–16. Oxford and New York: Oxford University Press.

Tarazona, Ariel M., Maria C. Ceballos, and Donald M. Broom. 2019. Human relationships with domestic and other animals: One Health, One Welfare, One Biology. *Animals* 10:43.

Van Patter, Lauren E., Julia Linares-Roake, and Andrea V. Breen. 2023. What does One Health want? Feminist, posthuman, and anti-colonial possibilities. *One Health Outlook* 5:4.

Waltner-Toews, David. 2009. Eco-Health: A primer for veterinarians. *Canadian Veterinary Journal* 50 (5): 519–521.

Weinhold, Bob. 2003. Conservation Medicine: Combining the best of all worlds. *Environmental Health Perspectives* 111 (10): A525–A529.

Whitmee, Sarah, Andy Haines, Chris Beyrer, Frederick Boltz, Anthony G. Capon, Braulio Ferreira de Souza Dias, Alex Ezeh, Howard Frumkin, Peng Gong, Peter Head, Richard Horton, Georgina M. Mace, Robert Marten, Samuel S. Myers, Sania Nishtar, Steven A. Osofsky, Subhrendu K. Pattanayak, Montira J. Pongsiri, Cristina Romanelli, Agnes Soucat, Jeanette Vega, and Derek Yach. 2015. Safeguarding human health in the Anthropocene epoch: Report of The Rockefeller Foundation-Lancet Commission on planetary health. *Lancet* 386:1973–2028.

WHO. 2021. *Tripartite and UNEP support OHHLEP's definition of "One Health"*. Joint Tripartite (FAO, OIE, WHO) and UNEP Statement. WHO Joint News Release December 1, 2021.

Wilcox, Bruce A., A. Alonso Aguirre, Peter Daszak, Pierre Horwitz, Pim Martens, Margot Parkes, Jonathan A. Patz, and David Waltner-Toews. 2004. EcoHealth: A transdisciplinary imperative for a sustainable future. *EcoHealth* 1:3–5.

Williamson, L. E., C. A. Dell, D. Chalmers, M. Cruz, and P. De Groot. 2022. Strengthening Zooeyia: Understanding the human-animal bond between veterans living with comorbid substance use and posttraumatic stress disorder and their service dogs. *Human Animal Interaction Bulletin* 10 (2): 20–47.

Zinsstag, Jakob, Martyn Jeggo, Esther Schelling, Bassirou Bonfoh, David Waltner-Toews, Stefano Lelii, Bob Williams, Mahamat Bechir, Felix Li, Dominique Charron, Robinson Mdegela, Meredith Barrett, Richard Kock, Ibrahima Sy, Alain Vandersmissen, Peter Black, and Peter Daszak. 2012. Convergence of EcoHealth and One Health. *EcoHealth* 9:371–373.

Zinsstag, Jakob, David Waltner-Toews, and Marcel Tanner. 2021. Why One Health? In One Health: The theory and practice of integrated health approaches, ed. Zinsstag, J, E. Schelling, L. Crump, M. Whittaker, M. Tanner, and C. Stephen, 15–24. 2nd ed. Wallingford: CABI.

Chapter 3
Ethical Attempts for One Health Approaches

> *...everywhere the two medical services of the Beloved of the Gods, ..., have been provided. These consist of the medical care of man and the care of animals. Medicinal herbs whether useful to man or to beast, have been brought and planted wherever they did not grow; similarly, roots and fruit have been brought and planted wherever they did not grow. Along the roads wells have been dug and trees planted for the use of men and beasts (2nd major rock edict; Thapar 1997, 377.)*

Abstract This chapter reviews the still pristine ethical discussion within the different One Health approaches. Calls for incorporating ethics have been present in the approaches since approximately 2009. In this chapter, I will thematically present ethical frameworks proposed within scientific texts from the One Health approaches, during the period covered from 2004 up to June 2023. Therefore, this Chapter provides an overview of the field, along with brief analyses of suggested frameworks that could be useful for further exploration in upcoming chapters. The frameworks selected for the following chapters of this book were Principlism, Biophilia, Capability Approach, Four Principles of Animal Ethics, and the Land Ethic. Other ethical perspectives to consider in further chapters were ethics of care and feminism, pluralism, and acknowledgement of Indigenous peoples' value systems.

Keywords Ethical competency · Pluralism · Manhattan principles · Canmore Declaration · Berlin Principles of One Health · São Paolo Declaration on Planetary Health · OHHLEP

H. Lerner, *Ethics for One Health Approaches: A Roadmap for Future Directions*, Advancing Global Bioethics 22,
https://doi.org/10.1007/978-3-032-05239-1_3

3.1 Introduction

The epigraph is one of the oldest written sources (on rocks) that indicates an ethical rule that could have been used within today's One Health approaches, stemming from the reign of King Aśoka in Ancient India. A full description of the history of ethics that foregone One Health approaches is not the aim of this book. However, it is a gentle reminder that some thoughts presented here might be thoughts that now and then have been present through the history of humans.

In this chapter I will thematically present proposed ethical frameworks made within the One Health approaches. I have chosen from core scientific texts on ethics written from 2004 up to June 2023. The method follows that described in Sect. 2.1.1. Several researchers have found a lack of ethics within the One Health approaches. I will start this chapter by considering such calls for ethics and then present answers to these calls. These answers will be presented thematically for the One Health approach, where they first appear. When other approaches also suggest the same framework, I have added these answers to the presentation of the framework although this might seem like a distortion of the thematic structure. One might get the impression that ethics has been more called for and elaborated in One Health compared to the other approaches. Have in mind that some researchers use the term "One Health" as referring to all One Health approaches.

The aim of this chapter is to make an overview of ethical suggestions in this rather pristine field to be able to choose proposed frameworks to analyze in the following chapters in Chaps. 4–7 of this book. Therefore, this text expands in various directions in these other Chapters. You might see this chapter as an overview roadmap, introducing you to the landscape and main roads. In the forthcoming Chaps. 4–7 you will get finer details. The aim of the entire book is to find frameworks for One Health approaches, the division into approaches here is only for dispositional matters.

3.2 A Call for Ethics

To understand the current state of affairs in ethics of One Health approaches, one needs to have a brief sketch of the history. One of the most agreed starting points for One Health as an approach is the Manhattan principles from 2004 which recognized the inextricable connection between humans, animals and the ecosystem (Cook et al. 2004). The Manhattan Principles had a slight touch of ethics although seldom recognized. I will develop the ethical content in this cornerstone later on in the Sect. 3.3.1.1.

The first call for ethics came in 2009. This was a scientific paper where lessons from the avian influenza and the swine flu were discussed (Scoones and Forster

2009).[1] It addressed equity issues indicating that ethics might be an important part of the approaches. They raised important questions of distributive issues and emphasized the need to ensure that culling of animals should not risk local livelihoods that depended on those animals. Although equity is an ethical issue, the text did not refer to ethics but rather to economic or social aspects.

In my own paper (Lerner 2013) on philosophical roots to the One Medicine approach (a forerunner to the narrow version of One Health) I analyzed at least three areas that have a bearing on One Health ethics. These could be summarized as

1. Studying valuation of different disciplines in interdisciplinary or transdisciplinary approaches,
2. Analysis of normative aspects of concepts of health
3. Ethical issues in healthcare treatment concerning the relation of care and the goal of care. (Lerner 2013)

The third area consists also of at least two kinds of analyses; how one values humans, animals and the environment, and ethical issues in research ethics (Lerner 2013, 104–105). I claimed that there was an early focus in One Medicine on trying to discuss the similarity between humans and animals. This could imply that animals and humans have equal values. I argued that humans were regarded as more important and that there was a weak or strong human priority-position in animal ethics rather than a human dominion-position or an equal consideration of interest-position.

Further, I draw attention to the possibilities of having similar ethical frameworks in veterinary medicine and human medicine which could work at least for what is now called a narrow version of One Health. Building on the work of Mullan and Main (2001), who argued that the biomedical principles proposed by Beauchamp and Childress—known as *Principlism*—used in human medicine could also be applied in veterinary medicine. At least in the narrow version of One Health, this could be a promising starting point for a discussion of ethics, and I will elaborate more on this suggestion in Chap. 4.

The claim that one could start with bioethics from the public health and healthcare sector as a starting point for ethical development has also been reinforced later by Degeling et al. (2019). However, they note that the anthropocentric focus of public health ethics might be troublesome for One Health approaches due to the recognition of animals and ecosystems as valuable.

> One Health policies and practices requires us to balance the needs of human individuals and populations with the health and wellbeing of non-human others and our shared ecological systems. (Degeling et al. 2019, 77)

Another early attempt was Lajaunie et al. (2014) stating the importance that One Health research considers research ethics within the wider version of bioethics such as the intention in the Convention of Biological Diversity (CBD) and the Nagoya Protocol. The Nagoya protocol has its focus on fair and equitable sharing regarding

---

[1] See also Sect. 3.4.1 for another early paper in 2009.

genetic resources and is based on Article 15 in the CBD and recognizes the importance of health preparedness on human pathogens which makes it linked to One Health approaches. One central aspect is the inclusion of indigenous and local communities with a clearly statement of them as those choosing the rightful holders of their knowledge (Secretariat of the Convention on Biological Diversity 2011). Other researchers claim the importance of encompassing an anti-colonial perspective in One Health approaches (Van Patter et al. 2023).[2] This will be further analyzed in Chap. 7 in this book.

Marcel Verweij and Bernice Bovenkerk suggested a starting point for the ethical research in One Health approaches by encouraging researchers to start doing work and study assumptions in the scientific, metaphysical, conceptual, and moral areas (Verweij and Bovenkerk 2016). By having a broad philosophical approach, one minimizes the risk of missing aspects crucial for ethics.

3.2.1 Call for Pluralism

There has also been a call for pluralistic ethical frameworks. In 2015, Paul B. Thompson and Monica List (2015) called for "One Bioethics" which was a supposed merge of medical bioethics and environmental philosophy. This call was the result of analyzing an outbreak of Ebola in 2014. One of the main interesting points in the paper is a claim by H. Brody which is cited as:

> …new directions for bioethics, including cross-cultural ethical relativity and global-scale resource allocation for the prevention and mitigation of international health crises. … prioritize topics that have been historically neglected by the field. (Thompson and List 2015, 101)

Their point is that there needs to be a focus on bridging ethical differences, such as crosscultural ethical relativity but also the divide between anthropocentrism and bio-/ecocentrism. This could be seen as a call for pluralism in ethics. However, the term "One Bioethics" has rarely been used, and the paper merely introduce the concept without going in depth on its demarcation. The plurality idea has also interested other scholars. Chris Degeling et al. elaborates more modestly on pluralism.

> …a modest form of pluralism (committed only to the view that there is more than one morally important value) (Degeling et al. 2019, 74)

This statement is shallow, and one might question if it describes pluralism at all. This is only a pluralism in values within one and the same decision-making theory. Lauren E. Van Patter et al. (2023) argue for a broad pluralism that encompass almost everything originating from feminism (ethics of care) and ecofeminism:

---

[2] Van Patter et al. (2023) proposes three perspectives that One Health approaches need to consider, anti-colonialism, multispecies solidarity and justice (within posthumanism), and feminist care ethics. Feminist care ethics will be briefly mentioned further down in this Chapter (and more elaborated in Chap. 7). Multispecies solidarity and justice will be mentioned in Sect. 3.5.1.2.

> The ethics of care theoretical framework creates space for inter- and intra-species collaboration, with reflection on moral quandaries that come with balancing human, animal, and environmental health. (Van Patter et al. 2023, 4)

I will discuss aspects of pluralism at length in Chap. 7. Now I will turn to proposals from the different One Health approaches in the same order as they have been presented in Chap. 2.

3.3 Proposals from One Health

There is a slight predominance of ethical proposals that stem from scientific contributions within One Health. This might give the impression that this approach has emphasized ethics more than the other approaches. I think this is a bias, mainly due to the double use of the term One Health, which could be both referring to a particular approach, the One Health approach, or to all One Health approaches (see Sect. 1.1). Researchers might use One Health as a shortform for multi-species, interdisciplinary approaches opposite to global public health.

Two kinds of proposals exist in One Health, attempts to formulate principles for the approach (such as a declaration of intention) and suggestions for ethical frameworks. I will present the first in Sect. 3.3.1 and the latter in 3.3.2.

3.3.1 Principles for the Approach

There have been three important documents that could be seen as declarations within One Health, which also at least say something about ethics. First, I will consider the cornerstone for the One Health approach, The Manhattan Principles (Cook et al. 2004). This document formed the transition from One Medicine, focusing on human medicine and veterinary medicine, to One Health, also including ecology. Several years later, the Manhattan principles were revisited in an attempt to make a more modern version. That attempt resulted in The Berlin Principles of One Health (Gruetzmacher et al. 2021), which will be the second document here. Third, the WHO *One Health High-Level Expert Panel* (OHHLEP et al. 2022) has proposed an ethical framework for One Health.

3.3.1.1 The Manhattan Principles

The Manhattan Principles is often considered the start of the modern wide One Health approach (see Sect. 2.3). They were developed in 2004 and linked human and animals to ecosystems (Cook et al. 2004). The only principle among the 12 principles addressing ethics is number 8:

> 8. Restrict the mass culling of free-ranging wildlife species for disease control to situations where there is a multidisciplinary, international scientific consensus that a wildlife population poses an urgent, significant threat to human health, food security, or wildlife health more broadly. (Cook et al. 2004)

This quote restricts how to cull wildlife species to prevent disease transmission. There needs to be a very strong consensus to allow culling of free-ranging wildlife species, and only when there is an urgent and significant threat.

3.3.1.2 The Berlin Principles of One Health

The Berlin Principles of One Health is an update of the Manhattan Principles (Gruetzmacher et al. 2021). In an initiative from the Climate and Environmental Foreign Policy Division at the German Federal Foreign Office together with the Wildlife Conservation Society called One Planet, One Health, One Future 200 participants from 47 countries joined in establishing new guiding principles 15 years later. The Berlin Principles has an ethical foundation based on the framework of the early One Health ethicists. This framework was further developed with work from Lysaght et al. (2017) adding the concepts of solidarity and environmental justice.

In the Berlin Principles, the ten principles are outlined under the heading *Towards health for all*. However, none of the principles presents explicitly any ethical guidance more than; one should foster inter- and transdisciplinarity, all stakeholders should be considered including indigenous peoples and local communities, and there is a essential link between humans, animals, plants and ecosystems recognizing human animal and environmental health. As I see it, this could be interpreted as meeting both the claims for solidarity and environmental justice but still the understanding of its implications remains shallow. Even if the conclusion in the presenting paper states that the "quality of current and future human and animal health and well-being depend on humanity's respectful, humble and responsible environmental stewardship" (Gruetzmacher et al. 2021), it provides little insight into how this can be achieved. The concepts and principles are not developed within the text.

3.3.1.3 One Health High-Level Expert Panel

In 2021, WHO initiated the One Health High-Level Expert Panel (OHHLEP), a body that will advise the WHO and three other international organizations; the FAO, UNEP, and the WOAH. The panel has already presented a kind of ethical framework consisting of a number of principles:

1. **Equity** between sectors and disciplines.
2. Sociopolitical and multicultural **parity** (the doctrine that all people are equal and deserve equal rights and opportunities) and inclusion and engagement of communities and marginalized voices.

3. Socio-ecological **equilibrium** that seeks a harmonious balance between human-animal-environment interaction and acknowledging the importance of biodiversity, access to sufficient natural space and resources, and the intrinsic value of all living things within the ecosystem.
4. **Stewardship** and the responsibility of humans to change behaviour and adopt sustainable solutions that recognize the importance of animal welfare and the integrity of the whole ecosystem, thus securing the wellbeing of current and future generations; and.
5. **Transdisciplinarity** and multisectoral collaboration, which includes all relevant disciplines, both modern and traditional forms of knowledge and a broad representative array of perspectives (OHHLEP et al. 2022, 3, original emphasis).

Here one can stress several ethical aspects. Equity is a core value and should exist both between sectors and disciplines within the approach avoiding marginalizing perspectives. Marginalized voices will be included when one acknowledges multicultural and sociopolitical parity and where both present and future generations are concerned. In my view, this implies a pluralistic ethics although not explicitly written in the document.

OHHLEP stated that all living things have intrinsic value and that integrity is applied to ecosystems. These are strong claims ruling out several possible anthropocentric and zoocentric ethical frameworks. Compared to strands within environmental ethics the text clearly asserts that the human attitude to nature belongs to the tradition of stewardship.

3.3.2 Towards an Ethical Framework

Two doctoral theses with a primary focus on ethics in One Health approaches exist (Lederman 2018; van Herten 2021). While van Herten ended up in a precaution-as-prevention-principle to work with animal culling (which is too narrow as an ethical guideline for all One Health approaches), Lederman studying the same issue, tried to construct an ethical framework, which is worth analyzing here.

The doctoral thesis of Zohar Lederman (2018) analyzed critically meta-ethical theories in environmental ethics. After an in-depth analysis of different criteria for moral status, Lederman chose to build his theory on biocentric meta-ethics. For him, the criterion for having moral status is being alive, and he follows the tradition with origins from Albert Schweitzer.

He ended up suggesting a procedural account of ethics with different steps to choose from among conflicting interest claims. Some actions are immediately unjustified if they violate the criteria of moral status. Based on David Hume and Ezekiel Emanuel a communitarian deliberation model was suggested with a public and inter-personal debate based on both sentiments and reason. Material such as public policies and principles from environmental ethics and public health ethics will be deliberated to create new public policies based on ethics. As a case through the thesis, he used the culling of badgers in England. He found culling unjustified both ethically and scientifically.

The biocentric idea of Schweitzer that all living forms have value has also been adopted quite early by Michael W. Fox (2011) in his *Healing Animals & The Vision of One Health*. This early book of ethics in the field of One Health ends up in a suggestion of his view of *equalitarianism*

> Such equalitarianism is the guiding principle of a new planetary politics based on an egalitarian intercultural and interspecies democracy, where we extend the Golden Rule to all beings. To have equal respect and concern for Nature and all living beings, and to live by the all-embracing Golden Rule, is ultimately enlightened self-interest. (Fox 2011, 185)

This book has been accused of lacking ethical rigor to be useable (Lederman 2018) and the call for extending the Golden Rule, to treat others as you want them to treat you, might be too shallow for an ethical framework of One Health approaches. Although the Golden Rule exists in many religions and could therefore be easily adopted, one must place the Golden Rule within a more rigid ethical framework, which I do not intend to do in this book. So, I dismiss Fox ethical suggestion. I also dismiss Lederman's theory for One Health approaches as I am not convinced that his argumentation is convincing why these theories could be combined and the conclusions he draws from the theory.

However, Lederman will be important throughout this book. The claim for an ethic that gives moral status to all living beings, as evident in both Lederman's and Fox's contributions indicates that one must regard all value perspectives, including bio- and ecocentrism, in One Health. Lederman (2018) takes this one step further. By analyzing definitions of One Health, he argues strongly that inherent in important definitions of the field is an ethical imperative that One Health promotes the health of humans, animals and ecosystems, and therefore also needs to give them value and include them in the ethical analysis. This ethical imperative of One Health approaches will influence parts om my analysis in Chaps. 4–7, as it aligns with much of the work I present in Chap. 2 regarding all the approaches.

3.4 Proposals from EcoHealth

Two different ethical frameworks, Biophilia and the Land ethic, have first been suggested in EcoHealth and then also in other One Health approaches. These suggestions are described below.

3.4.1 *Biophilia*

In EcoHealth there has been a suggestion that one needs a theory such as *Biophilia* dealing with nature and human relations to other species. Biophilia are developed by biologists Edward O. Wilson and Stephen R. Kellert (Maller et al. 2009). The core of this theory is that humans have an innate affiliation for valuing other species and the relations to other species. A close look at the core sources from Wilson and Kellert clarifies that the theory is primarily anthropocentric and has its focus on how humans value species and ecosystems. Biophilia has also later been suggested for One Health (Chalmers and Dell 2015). The value of this theory for One Health approaches has also been contested (Kingsley and Lawson 2015), and I will analyze Biophilia in Chap. 4.

3.4.2 *Land Ethic*

Within EcoHealth, Berkes et al. (2012) claimed that Aldo Leopold's concept of land health and the ethical implications of it in interconnectedness with humans are central for the approach. Later, Goldberg and Patz (2015) argue that global health strategies such as One Health and Planetary Health should adopt an ethical framework based on the thoughts of Aldo Leopold's Land ethic. The argument in favor is that the scientific basis in One Health and Planetary Health is based on interconnectedness which also is the foundation for the Land ethic. Their claim is not for a principle-based ethics for One Health approaches, but rather a procedural ethic with the value frame of recognizing the environment as valuable as humans. The resulting ethical guidelines will be developed in the "social evolution" in the societal debate according to Goldberg and Patz. Also, Radcliffe and Jessup (2022) discuss the importance of Leopold's thinking about wildlife health in the applications of wildlife conservation in North America, One Health, and Planetary Health. They also elaborate on his ethics and lifting core aspects to consider. I will further discuss Leopold's Land ethic in Chap. 6.

3.5 Proposals from Planetary Health

Two kinds of proposals exist in Planetary Health: attempts to formulate principles for the approach (such as a declaration of intention, Sect. 3.5.1) and suggestions for the ethical frameworks that Planetary Health should encompass (the latter presented in Sect. 3.5.2).

3.5.1 *Principles for the Approach*

The attempts to formulate principles for the approach consist of two declarations. The first is The Canmore Declaration (Sect. 3.5.1.1) and the second is the Sao Paolo Declaration (Sect. 3.5.1.2).

3.5.1.1 The Canmore Declaration

The Canmore Declaration is an attempt within the Planetary Health approach (Prescott et al. 2018) to show the inextricable connection between "health of people, places and the planet". Based on the Ottawa Charter, mainly with an anthropocentrical language, it also considers the connection between humans and manmade as well as natural environments. The Canmore Declaration consists of ten principles where ethical aspects are included within some of the principles. In principle 2 there is a description of how humans should be, namely have

> … human kindness, empathy, mutualism, responsibility, and reciprocity at the individual, community, societal and global levels … (Prescott et al. 2018, 3)

Injustice due to oppression from different systems in society should be avoided, as claimed in Principle 5. In Principle 6 they acknowledge that there are emotional bonds between nature, biodiversity, and people, and in Principle 7 that we as humans are dependent on the natural environment. In the final two principles they want to avoid elitism and social domination in society as well as proposing new normative behaviors for us as individuals. These normative behaviors are summarized as being a role model and

> … remain committed to a planetary peace agenda; encourage mutualism, empathy and community cohesion; and underscore that aggression, conflict and violence are destructive to person, place and planet. (Prescott et al. 2018, 3)

Although acknowledging the connection to nature, the Canmore Declaration solely has its focus on ethics for humans, how they should behave towards each other and partly towards nature. These statements could therefore be analyzed as proposed virtues. Empathy and freedom from oppression especially for marginalized groups are two crucial aspects of ethics, which might indicate influence from feminist ethics, but there is no firm presentation of the ethical foundation within the declaration. I will come back to feminist ethics in Chap. 7.

3.5.1.2 The São Paulo Declaration on Planetary Health

One of the recent declarations is the São Paulo Declaration on Planetary Health signed by the global planetary health community including over 250 organization signatories and were published October sixth, 2021 (Myers et al. 2021).[3]

Main ethical focus in the São Paulo Declaration is on justice and fairness. Its hope is to move the focus from an anthropocentric view of the world and the sustainability concept towards more inclusion. Instead of "exceptionalism, domination, and scarcity" one will aim for "interdependence, equity, and regeneration". The goal is to "safeguard a healthy and equitable future for humanity and protect all life on earth". Sustainability relies on the well-being of nature and society and rights are ascribed to both humans and the environment, although a specification is absent. Regeneration refers to the regeneration of nature and natural systems and is an important value to support. However, both the concept of health and the concept of equity are framed in a distinctly anthropocentric way, with a focus on humans, particularly in relation to issues like inclusion. Much of the text concerns particularly human well-being and human equity and inclusion.

Although briefly explained, the way of inclusion of marginalized groups as well as inclusion within research, policy making, and work is in my opinion a call for an ethic belonging to the field of discourse ethics, which will be further developed in Chap. 7. In that Chapter I have chosen the version proposed by Seyla Benhabib which I regard suits One Health approaches best. The motivation behind my choice will be explained in Chap. 7.

In my view, the declaration clearly strengthens the road of moving away from Planetary Health solely being an extended version of global or public health where an anthropocentric ethic would be favored towards something else. Environmental ethics as environmental rights are confirmed to be important and a perspective of human dominance of nature is abandoned. However, the text has too much emphasis on humans and there are still sentences written indicating that a healthy nature is a prerequisite for human flourishing.

One option would be to incorporate the concepts of multispecies solidarity and justice as guiding the discussion of ethics. This was proposed the same year in Planetary Health (Maric and Nicholls 2021) and later for One Health (Van Patter et al. 2023). I see multispecies solidarity and justice as a part of allowing other species to be part of the moral community, which implies that aspects of these two will be discussed in Chaps. 5 and 6 in the framework of other ethical theories.

---

[3] Published in Lancet October 9, 2021.

3.5.2 Towards an Ethical Framework

In 2020, Alexander Foster et al. proposed five ethical positions for Planetary Health. These could be seen as the beginning of an ethical framework emerging from within the approach. To guide their five ethical positions, they provided four criteria for ethics, referred to as foundations. These are

1. Ethical statements must be useful
2. Ethical statements must be usable
3. Ethical statements must be used
4. Ethical statements must be universal (Foster et al. 2020, 456)

Useful is that the statement must be clear enough to help solving dilemmas. *Usable* refers to how statements are framed; a statement should be both accessible and easy to work with. The phrase "Ethical statements must be *used*" might seem unnecessary for an ethicist as well as a body already working within an ethical framework. However, in these approaches, where the level of ethical thinking largely focuses on calling for ethics, it is important to explicitly state this. *Universality* is a core aspect of ethics shared by almost all ethical theories. An intrinsic value, such as human rights, must be universally recognized.

The five ethical positions based on the four criteria were intergenerational responsibility, extending rights to the more-than-human world, distributive justice, the precautionary principle, and the right to know (Foster et al. 2020, 457–466).

Intergenerational responsibility is by the authors claimed as an axiomatic principle within Planetary Health. Both deceased and forthcoming generations are considered (to be culturally and religious inclusive). They hold a stewardship view on humans, a position that also has been argued for by Nicole de Paula:

> ...human stewardship of the Earth is a primary determinant of future population health. (de Paula 2021, 142)
>
> Without intergenerational solidarity and a genuine drive to include the Global South's voices and minorities in the North, the vision of planetary health will not be achievable by any definition. (de Paula 2021, 13)

The risk that the principle is too abstract to be useful could be solved by defining the number of generations. Foster et al. (2020) claims that it may be possible to grasp and consider three generations back and three generations forward in time.

Extending rights to the more-than-human world[4] implies valuing species or ecosystems as something more than instrumental for humans. Often, intrinsic values are used for this, but Foster et al. (2020) lifts the development of a third category, relational values, which has been at the fore in the work of Intergovernmental Science-Policy Platform on Biodiversity and Ecosystem Services (IPBES). I will return to this in Sect. 6.2.1.

---

[4]This way of referring to other entities than humans in ethics has also been used in an edited book by Irus Braverman (2023), with phrases such as "more-than-One Health", "more-than-human", and "more-than-almond" to mention a few within the different chapters' perspectives.

Distributive justice must be understood as how to move towards a society that has more equity. According to Foster et al. there are three topics that are particularly important within Planetary Health. These are:

- Inequalities in who has contributed to planetary degradation;
- Inequalities in who is most harmed by planetary degradation; and.
- Opportunities to rectify injustices while addressing planetary degradation (Foster et al. 2020, 459).

Issues needed to be addressed are burdens of colonialism but also claims that ethical frameworks need to be anti-colonial (van Patter et al. 2023). This could be linked to intergenerational responsibility (see quote from de Paula above) with specific emphasis to inequities held by Indigenous peoples and the global south.

The precautionary principle is similar as mentioned in the presentation of suggestions from the One Health approach in Sect. 3.3 although in Planetary Health focus is on environmental aspects.

The right to know is the right to be aware of the risks in order to make informed decisions and the way of presenting the risks should be done by a dialogue (Foster et al. 2020). Foster et al. present several aspects that facilitate this process which in a condensate quote becomes:

1. Information must be presented honestly and objectively
2. The duty to communicate risk is a duty both to listen attentively and to communicate through the language and worldview of the affected people
3. A right to know is also a right to hope (Foster et al. 2020, 465–466).

When I consider these aspects along with text passages I have omitted in this quote, I get the impression of a one-sided monologue. Information transfer from one informing part to another affected part seems to be crucial (Foster et al. 2020).

De Paula (2021) argues that interpersonal communication connected with trust is core for reaching deliberations. For a successful Planetary Health approach, she envisages that it needs to be built on inclusivity, trust, and cooperation. Inclusivity is in order to share responsibility and power more equally. Perspectives that need to be included, but are often left out, are gender (as she states—women need to be at the discussion table and in deliberations a gender expert is required, indicating that de Paula relies on feminist ethics) and Indigenous knowledge. People involved in facilitating this interpersonal communication need several abilities that could be seen as virtues (although she doesn't label them as such).

> However, to navigate complex systems, as Planetary Health requires, specific characteristics must be accentuated. These include curiosity, humility, acceptance of uncertainty, reflexivity (being conscious of one's own role, prejudices and power), and being open to different ways of seeing the world. (de Paula 2021, 84)

De Paula (2021) mentions curiosity, humility, acceptance of uncertainty, reflexivity, and openness to different perspectives of the world. She also sees learning, loving, and truth-telling as core, and specifically stresses the importance of courage and understanding what courage should be.

> Contrary to assumption, courage is not about showing strength. It is, rather, about being comfortable with our inevitable vulnerabilities. (de Paula 2021, p. 161)

Another way of facilitating trust in interpersonal communication is advocating the possibility of safe space, where vulnerable groups can be heard and get an opportunity to reframe the discussion from their point of view (de Paula 2021, 95). I will return to dialogue ethics in Chap. 7.

3.6 Proposals from One Welfare

Much emphasis on ethics in One Health approaches has been on health and survival primarily (culling of animals, eradication of disease). In 2018, One Welfare, with a foundation in ethical issues with focus on mental health and well-being in humans and animals emerged from the One Health approach (García Pinillos 2018).

The One Welfare approach relies on the concepts of animal welfare and human well-being has primarily focus on mental health and well-being of animals and humans. Minimising abuse and suffering are core aspects here. In the One Welfare approach, at least five central areas for ethical concern are evident, here quoted from García Pinillos:

> The connections between animal and human abuse and neglect.
> The social implications of improved animal welfare.
> Animal health and welfare, human well-being, food security and sustainability.
> Assisted interventions involving animals, humans and the environment.
> Sustainability: connections between biodiversity, the environment, animal welfare and human well-being. (García Pinillos 2018, 14)

Although based on values and ethical issues linked to well-being and welfare, no particular ethical theory is explicit within One Welfare. There have been a few suggestions for ethical frameworks that could be suitable to use within One Welfare. These will be mentioned here and developed later in this book.

Anne Fawcett et al. (2018) suggested that a principle-based animal ethic consisting of four principles, a kind of principlism developed by David Fraser, would be suitable within the One Welfare approach. According to their study Fraser's principlism bridge a gap between animal ethics and environmental ethics, at least in the area of conservation. This will be further explained in Chap. 5, when this ethical framework will be analyzed.

Henrik Lerner (2019) stated that Martha Nussbaum's capability approach would be suitable for One Welfare, at least in animal assisted interventions. The capability approach will be presented and analyzed in Chap. 5.

3.7 Ethical Competency

There has also been a discussion on ethical competency. In 2011 we (Lerner et al. 2011) analyzed through empirical ethics how veterinarians, with their expert knowledge on suffering in their patients as well as their patient's relatives, i.e. their owners, could give insights in the research on attitudes to physician assisted suicide (PAS) in humans. This could be seen as a narrow One Health research approach, comparing veterinary medicine and human medicine with a research group consisting on veterinary, medical and ethical expertise. We had the assumption that veterinarians that were experts in alleviating suffering, sometimes with euthanasia, could have ethical competency even in this issue for humans. They could reach a different solution on the reasons to argue for and against PAS compared to the physicians which work with prolonging life. Three groups, physicians, lay people, and veterinarians got the same questionnaire asking for if they were positive or negative to PAS and how they ranked ethical arguments for and against PAS. The survey of the veterinarians was compared to similar surveys of physicians and lay people. Where physicians were negative to allowing PAS, veterinarians and lay people were positive. The analysis clarified that the veterinarians and the lay people differed in their argumentation for pro-PAS. Veterinarians based their arguments on their expert knowledge gained through their discipline. Part of the discussion in the paper was also on similarities on the triad of physician/veterinarian, patient and relative which is apparent in both veterinary medicine and human medicine.

Today, there are more structured analyses of ethical competency within One Health approaches. Quite recently, in a paper by the Network for Ecohealth and One Health (NEOH), they elaborated on core competencies for persons working within One Health approaches (Laing et al. 2023).

Within this framework ethics is acknowledged in one of their nine core competencies, namely within the competence of *social, cultural and gender equity and inclusiveness*. Their description of that competency is that one

> Understands and is sensitive to equity and justice in human terms and across animal and plant species and the environment. Acknowledges and incorporates gender and culture considerations. Shows integrity, behaving in a honest, fair and ethical manner with inclusiveness and humility. Acknowledges that we are part of nature (Laing et al. 2023, 7)

The same authors argues that another competency—effective communication—needs a dialogical form that bring

> ...social, relational and cultural contexts into their dialogue and co-generate and negotiate meaning. (Laing et al. 2023, 6)

They also recognize that the dialogue must bridge different knowledge and value systems and be aware of risks of

> disruption [that] occurs through language, physical, emotional or perceptional, cultural, interpersonal and gender barriers (Laing et al. 2023, 6)

To minimize inequities based on power relations. This implies, similar to the claim by Lederman, that one needs to bridge anthropocentrism, zoocentrism and bio-/ecocentrism. Laing et al. also propose at least rudimentary guidelines for a forthcoming procedural ethic. Some of the barriers for ethics that they recognize and need to be considered in such a procedural ethic is the following:

- A human-centred approach not recognizing that also animals, plants and environment have their own right to health.
- Gender and cultural discrimination.
- Society as a homogenous group, excluding people with differences.
- Focus on novel technologies despite these might be driven by unequal social drivers (Laing et al. 2023, 7).

The procedural ethic needs to avoid these barriers by an inclusive approach. Gender and cultural discrimination have been issues with too little emphasis in the literature of the One Health approaches (Friedson-Ridenour et al. 2019; van Patter et al. 2023). I will return to these issues, especially in Chap. 7.

3.8 Progress or Circular Reasoning?

Has there been any progress in the field? Both yes and no. Evidently, several ethical frameworks have been proposed which indicate progress. Still, it seems like an in-depth analysis is still awaiting and ethics seems still to be knocking on the door to be invited within the core sciences of One Health approaches. Quite recently, Joann M. Lindenmayer et al. again calls for ethics, providing four tenets to facilitate finding ethics.

1. One Health recognizes the interdependence of health and well-being among people, other animals, and nature.
2. One Health incorporates physiological, mental, emotional, and social well-being regardless of species, as well as the physical systems and ecological integrity of nature.
3. One Health strives to achieve optimal health simultaneously for humans, other animals, and nature, and accommodates tradeoffs that may not privilege only humans.
4. One Health acknowledges scientific uncertainty and our limited understanding of a complex and constantly changing world. (Lindenmayer et al. 2022, 3).

These four tenets set a similar framework to the one I have already established for One Health in Chap. 2 in this book. An ethic for One Health needs to consider health as a broad concept regarding aspects and species. It could not be fully anthropocentric and there are complex and changing issues to solve where there is also scientific uncertainty present, which implies that one must have an open system able to deal with new findings. This partly establishes the scientific normative framework while also outlining which values concerned as crucial.

In a paper by Coghlan et al. (2021), they claim that One Health needs to loose the anthropocentrical focus and expand ethics especially to non-human animals and also into biocentrism and ecocentrism, in what one of the co-authors, Peter Singer, calls the expanding circle. Besides the expansion to animals and ecosystems, they also call for more discussion on equity issues in global human health. However, this paper is just a call for more ethics and not a proposal of an ethical framework.

Has ethics been established as an important contribution to One Health approaches? A very brief look at two scientific roadmaps for the future One Health approaches indicates that one (Destoumieux-Garzón et al. 2018) consider a wide bioethics approach as crucial, while the other (Queenan et al. 2017) doesn't include ethics as a field. The latter, despite recognizing equity as an important ethical value, avoids recognizing this development. This is similar to the recent paper by Lindenmayer et al. (2022) quoted in this section, that still only calls for ethics without establishing it. The time is ripe for a more in-depth ethical discussion.

3.9 Summary

In this Chapter I have thematically structured proposed ethical frameworks within the One Health approaches. Despite my ambitious research (as explained in 2.1.1), I might have missed some proposals. As I have presented here, there are at least enough proposals to perform an in-depth analysis of the field.

I will give some concluding remarks while presenting how Part 2 (Chaps. 4–7) of the book will develop.

Crucial for values within One Health approaches is whether one primarily sees humans and animals (Lerner 2013), humans and environment (Goldberg and Patz 2015), or humans, animals and environment on par (OHHLEP et al. 2022). The ethical imperative argued for by Lederman (2018) is that One Health approaches need to choose the latter, humans, animals and environment on par. Because I am interested in what the widest version of One Health ethics could look like, I am inclined to follow this imperative.

The next step is now to start the in-depth analysis of proposed ethics within One Health approaches. This task will follow in the continuing chapters of this book. In Table 3.1 I summarize the different ethical frameworks presented in this Chapter with the chapter where I will primarily discuss them in the rest of the book.

In Chap. 4 I will analyze the suggestions for Principlism and Biophilia. Chapter 5 will analyze the Capability approach as well as the Four principles of animal ethics brought forward within One Welfare. In Chap. 6 the Land ethic will be discussed. I will also add a discussion on intergovernmental work on relational values as a specific kind of value besides the instrumental and intrinsic value in line with suggestions from Planetary Health.

Table 3.1 This chapter's proposed ethical frameworks with references to the chapters where they will be further discussed

Ethical framework	Approach suggested for	Source for suggestion	Analyzed in chapter
Principlism	One Health	Lerner (2013)	4
Biophilia	One Health EcoHealth	Maller et al. (2009) Chalmers and Dell (2015)	4
Capability approach	One Welfare	Lerner (2019)	5
Four principles of animal ethics	One Welfare	Fawcett et al. (2018)	5
Land ethic	One Health EcoHealth Planetary Health	Berkes et al. (2012) Goldberg and Patz (2015)	6
Ethics of care and Ecofeminism	Planetary Health	Van Patter et al. (2023)	7
Indigenous peoples valuesystem	One Health	Lajaunie et al. (2014)	7
Ethical pluralism	All	Thompson and List (2015) Degeling et al. (2019) Van Patter et al. (2023)	7

Chapter 7 will analyze pluralistic aspects such as Indigenous peoples claim on ethics as well as some insights from feminist ethics. The focus is on ethical dialogue, which will meet several of the claims made in this chapter. I will particularly introduce and discuss Seyla Benhabib's version which is cultural, gender and power sensitive. Here I will also add some thoughts on the ethical trend mentioned in Chap. 1 on discussions on individuality by considering the idea of Ubuntu and finally propose my own version of dialogue ethics that hopefully is gender-sensitive, non-colonial, and recognize multispecies solidarity.

References

Berkes, Fikret, Nancy C. Doubleday, and Graeme S. Cumming. 2012. Aldo Leopold's land health from a resilience point of view: Self-renewal capacity of social-ecological systems. *EcoHealth* 9:278–287.

Braverman, Irus, ed. 2023. *More-than-one health: Humans, animals, and the environment post-covid. routledge studies in environment and health*. London and New York: Routledge.

Chalmers, Darlene, and Colleen Anne Dell. 2015. Applying one health to the study of animal-assisted interventions. *EcoHealth* 12:560–562.

Coghlan, Simon, Benjamin John Coghlan, Anthony Capon, and Peter Singer. 2021. A bolder one health: Expanding the moral circle to optimize health for all. *One Health Outlook* 3:21.

Cook, Robert A., William B. Karesh, and Steven A. Osofsky. 2004. *The manhattan principles on "one world, one health"*. http://www.wcs-ahead.org/manhattan_principles.html

Degeling, Chris, Angus Dawson, and Gwendolyn L. Gilbert. 2019. The ethics of one health. In *One planet, one health*, ed. Merrilyn Walton, 65–84. Sydney: Sydney University Press.

de Paula, Nicole. 2021. *Breaking the silos for planetary health: A roadmap for a resilient post-pandemic world*. Singapore: Palgrave Macmillan.

Destoumieux-Garzón, Delphine, Patrick Mavingui, Gilles Boetsch, Jérome Boissier, Frédéric Darriet, Priscilla Duboz, Clémentine Fritsch, Patrick Giradoux, Frédérique Le Roux, Serge Morand, Christine Paillard, Dominique Pontier, Cédric Sueur, and Yann Voituron. 2018. The One Health concept: 10 years old and a long road ahead. *Frontiers in Veterinary Science* 5:14.

Fawcett, Anne, Siobhan Mullan, and Paul McGreevy. 2018. Application of Fraser's "practical" ethic in veterinary practice, and its compatibility with a "One Welfare" framework. *Animals* 8:109.

Foster, Alexander, Jennifer Cole, Ivica Petrikova, Andrew Farlow, and Howard Frumkin. 2020. Planetary health ethics. In *Planetary health: Protecting nature to protect ourselves*, ed. Samuel Myers and Howard Frumkin, 453–473. Washington and Covelo: Island Press.

Fox, Michael W. 2011. *Healing animals & the vision of one health*. Sarasota: One Health Vision Press.

Friedson-Ridenour, Sophia, Tracey V. Dutcher, Claudia Calderon, Lori Di Prete Brown, and Christopher W. Olsen. 2019. Gender analysis for one health: Theoretical perspectives and recommendations for practice. *EcoHealth* 16:306–316.

García Pinillos, Rebeca. 2018. *One welfare: A framework to improve animal welfare and human well-being*. CABI.

Goldberg, Tony L., and Jonathan A. Patz. 2015. The need for a global health ethic. *Lancet* 386:e37–e39. https://doi.org/10.1016/S0140-6736(15)60757-7.

Gruetzmacher, Kim, William B. Karesh, John H. Amuasi, Adnan Arshad, Andrew Far-low, Sabine Gabrysh, Jens Jetzkowitz, Susan Lieberman, Clare Palmer, Andrea S. Winkler, and Chris Walzer. 2021. The Berlin principles of one health: Bridging global health and conservation. *Science of the Total Environment* 764:142919.

van Herten, Joost. 2021. *Considerations for an ethic of one health: Towards a socially responsible zoonotic disease control*. Ph. D. thesis. Wageningen University.

Kingsley, Jonathan, and Justin Lawson. 2015. Finding a unified understanding of nature. *EcoHealth* 12:551–552.

Laing, Gabrielle, Eleanor Duffy, Neil Anderson, Nicolas Antoine-Moussiaux, Maurizio Aragrande, Caetano Luiz Beber, John Berezowski, Elena Boriani, Massimo Canali, Luis Pedro Carmo, Ilias Chantziaras, Glen Cousquer, Daniele De Meneghi, Ana Gloria Rodrigues Sanches da Fonseca, Julie Garnier, Martin Hitziger, Thomas Jaenisch, Hans Keune, Claire Lajaunie, Lorena Franco Martinez, Rebecca Maudling, Marie K. McIntyre, Barry J. McMahon, Alberto Munoz Prieto, Liza Rosenbaum Nielsen, Ranya Özçelik, John W.A. Rossen, Simon R. Rüegg, Sara Savić, Margarida Pires Simoes, Deborah J. Thomson, Laura Tomassone, Asta Tvarijonaviciute, Manuela Vilhena, Barbara Vogler, and Barbara Häsler. 2023. Advancing one health: Updated core competencies. *CABI One Health*. https://doi.org/10.1079/cabionehealth.2023.0002

Lajaunie, Claire, Serge Morand, and Tan Boon Huan. 2014. Barcoding, biobanking, ebanking for "One Health" projects in South-East Asia: Considering ethics and inter-national law. *Eubios Journal of Asian and International Bioethics* 24:129–131.

Lederman, Zohar. 2018. *Stamping out culling: The ethics of culling from a one health perspective*. Ph. D. Thesis. Centre for Biomedical Ethics, National University of Singapore.

Lerner, Henrik. 2013. The philosophical roots of the "One Medicine" movement: An analysis of some relevant ideas by Rudolf Virchow and Calvin Schwabe with their modern implications. *Studia Philosophica Estonica* 6 (2): 97–109.

Lerner, Henrik. 2019. A proposal for a comprehensive human-animal approach of evaluation for animal-assisted interventions. *International Journal of Environmental Research and Public Health* 16:4305.

Lerner, Henrik, Anna Lindblad, Bo Algers, and Niels Lynöe. 2011. Veterinary surgeons' attitudes towards physician-assisted suicide: An empirical study of Swedish experts on euthanasia. *Journal of Medical Ethics* 37:295–298.

Lindenmayer, Joann M., Gretchen E. Kaufman, Liv Baker, Simon Coghlan, Fred W. Koontz, Joachim Nieuwland, Kristin L. Stewart, and William S. Lynn. 2022. One Health ethics: What then must we do? *CABI One Health*. https://doi.org/10.1079/cabionehealth.2022.0011

Lysaght, Tamra, Benjamin Capps, Michele Bailey, David Bickford, Richard Coker, Zohar Lederman, Sangeetha Watson, and Paul Anantharajah Tambyah. 2017. Justice is the missing link in one health: Results of a mixed methods study in an urban city state. *PLoS One* 12 (1): e0170967.

Maller, Cecily J., Claire Henderson-Wilson, and Mardie Townsend. 2009. Rediscovering nature in everyday settings: Or how to create healthy environments and healthy people. *EcoHealth* 6:553–556.

Maric, Filip, and David A. Nicholls. 2021. Environmental physiotherapy and the case for multispecies justice in planetary health. *Physiotherapy Theory and Practice: An International Journal of Physical Therapy* 38 (13): 2295–2306.

Mullan, S., and D. Main. 2001. Principles of ethical decision-making in veterinary practice. *In Practice* 23:394–401.

Myers, Samuel S., Jeremy I. Pivor, and Antonio M. Saraiva. 2021. The São Paolo declaration on planetary health. *Lancet* 398:1299. including Supplementary appendix.

One Health High-Level Expert Panel (OHHLEP), Wiku B. Adisasmito, Salama Almuhairi, Casey Barton Behravesh, Pépé Bilivogui, Salome A. Bukachi, Natalia Casas, Natalia Cediel Becerra, Dominique F. Charron, Abhishek Chaudhary, Janice R. Ciacci Zanella, Andrew A. Cunningham, Osman Dar, Nitish Debnath, Baptiste Dungu, Elmoubasher Farag, George F. Gao, David T. S. Hayman, Margaret Khaitsa, Marion P. G. Koopmans, Catherine Machalaba, John S. Mackenzie, Wanda Markotter, Thomas C. Mettenleiter, Serge Morand, Vyacheslav Smolenskiy, and Lei Zhou. 2022. One health: A new definition for a sustainable and healthy future. *PLoS Pathogens* 18 (6): e1010537.

Prescott, Susan L., Alan C. Logan, Glenn Albrecht, Dianne E. Campbell, Julian Crane, Ashlee Cunsolo, John W. Holloway, Anita L. Kozyrskyj, Christopher A. Lowry, John Penders, Nicole Redvers, Harald Renz, Jakob Stokholm, Cecilie Svanes, Ganesa Wegienka, and on Behalf of inVIVO Planetary Health, of the Worldwide Universities Network (WUN). 2018. The Canmore Declaration: Statement of principles for Planetary Health. *Challenges* 9:31.

Queenan, Kevin, Julie Garnier, Liza Rosenbaum Nielsen, Sandra Buttigieg, Daniele de Meneghi, Martin Holmberg, Jakob Zinsstag, Simon Rüegg, Barbara Häsler, and Richard Kock. 2017. Roadmap to a one health agenda 2030. *CAB Reviews* 2017 12, No. 014.

Radcliffe, Robin W., and David A. Jessup. 2022. Wildlife health and the North American model of wildlife conservation. *Journal of Zoo and Wildlife Medicine* 53 (3): 493–503.

Scoones, Ian, and Paul Forster. 2009. One world, one health? *Rural* 21 (6): 22–24.

Secretariat of the Convention on Biological Diversity. 2011. Nagoya protocol on access to genetic resources and the fair and equitable sharing of benefits arising from their utilization to the convention on biological diversity: Text and annex. Secretariat of the Convention on Biological Diversity, United Nations, Montreal, 15 pages.

Thapar, Romila. 2012. *Aśoka and the decline of the Mauryas*. 3rd ed. Oxford: Oxford University Press.

Thompson, Paul B., and Monica List. 2015. Ebola needs one bioethics. *Ethics, Policy & Environment* 18 (1): 96–102.

Van Patter, Lauren E., Julia Linares-Roake, and Andrea V. Breen. 2023. What does one health want? Feminist, posthuman, and anticolonial possibilities. *One Health Outlook* 5:4.

Verweij, Marcel, and Bernice Bovenkerk. 2016. Ethical promises and pitfalls of onehealth. *Public Health Ethics* 9 (1): 1–4.

Part II
Possible Ethical Frameworks

Chapter 4
Expanding Anthropocentric Ethics

Whereas recognition of the inherent dignity and of the equal and inalienable rights of all members of the human family is the foundation of freedom, justice and peace in the world. (UN (1948), Preamble of Universal Declaration of Human Rights)

Abstract This Chapter is the first of the analyzing chapters of the proposals found within One Health approaches in Chaps. 2 and 3. Main emphasis of this chapter is suggestions of ethical frameworks originating from an anthropocentric view. After a brief introduction where I discuss the importance of agreeing on human rights, I will give an overview of anthropocentric bioethics. Thereafter, I analyze the two suggestions being made within One Health approaches, Principlism and Biophilia. Principlism, by Tom L. Beauchamp and James F. Childress, is an ethical framework that claims that regardless of ethical conviction, one can adopt the use of four ethical principles that will guide ethical decisions in the healthcare sector. Biophilia is a framework proposed by Edward O. Wilson and further developed by Stephen R. Kellert, which has its origin in ecology and describers values that humans have in relation to nature. Both these are dismissed due to the problem of not being inclusive enough for One Health approaches. They also violate the claim that One Health approaches need to consider humans, animals, plants, and ecosystems.

Keywords Principlism · Biophilia · Stephen R. Kellert · Edward O. Wilson · Tom Beauchamp · James Childress · Sustainability

H. Lerner, *Ethics for One Health Approaches: A Roadmap for Future Directions*, Advancing Global Bioethics 22,
https://doi.org/10.1007/978-3-032-05239-1_4

4.1 Introduction

In Chap. 2, the foundations of the different One Health approaches were presented. Chapter 3 summarized suggestions from One Health approaches about ethical frameworks needed. Both these empirical chapters gave several answers which could be differentiated into the anthropocentric ethics (anthropocentrism), animal ethics (zoocentrism) and environmental ethics (biocentrism and ecocentrism). In this chapter I will turn to normative ethics and start analyzing the suggestions originating from an anthropocentric view. After a very brief overview of anthropocentric bioethics and the ethical framework of human rights established by the United Nations, each main framework proposed, Principlism and Biophilia, will be presented and analyzed. The structure of the presentation will start in its original form and then it will be developed or criticized depending on applicability to One Health approaches. Finally, it will be applied to the three cases developed in Sect. 1.4. The results from this application will show strengths and weaknesses of these frameworks.

4.2 Overview of Anthropocentric Bioethics

The claim made for an exclusion of Planetary health from the One Health approaches is that this approach is no more than public health or global health expanded. No specific intrinsic value is attributed to either animals, plants or ecosystems. Following this line of argument, one might also claim that anthropocentric ethics is not fruitful at all for One Health approaches. Anthropocentric ethics has its primary focus on humans. Anthropocentrism is divided into at least two strands. There could be strong anthropocentrism where only humans count or weak anthropocentrism where non-human individuals also count but less than humans. In present public health debates concern might be for the environment and animals, but these are only instrumental to human health and well-being. Both weak and strong anthropocentrism fail to acknowledge the importance of animals, plants or ecosystems as being as important as humans.

Among the grandparents of One Health, in the writings of Rudolf Virchow and Calvin Schwabe, a weak anthropocentrism was evident indicating a possibility to reach an ethics for One Health from anthropocentrism (Lerner 2013). Also, if we disregard the claim that One Health approaches must value animals, plants or ecosystems on par with humans, a weak anthropocentric ethic might be plausible for at least Planetary Health.

Among the various frameworks of anthropocentric ethics two have been suggested within the One Health approaches (see Chap. 3), the four-principles approach to bio-medical ethics (or Principlism), and Biophilia.

Principlism originated in an American context and deals with ethical issues within a narrow version of bioethics, mostly human healthcare. It has gained influence at least in Western healthcare. Biophilia originated from ecology by the ecologist Edward O. Wilson. It has been further developed by Stephen R. Kellert. Before we discuss these, we need to consider the role and value of humans in international agreements such as the human rights and the sustainable development goals. One of the strongest arguments for having an emphasis on humans derives from the work with human rights (as referred to in the epigraph of this chapter) and the efforts to improve livelihoods for the most vulnerable groups in society. These vulnerable groups often share multiple burdens from pollution, environmental degradations, poverty etc.

4.2.1 United Nations and Humans

United Nations (UN), which by their organization the World Health Organization (WHO), is one of the strongest bodies that propose One Health approaches. Shortly after World War II the UN decided on the Human Rights . One of the human rights is the right to health and well-being, which is echoed in the sustainable development goal 3 (SDG 3), Good Health and Well-being (UN 2023). Health is understood as the WHO definition of health, which covers physical, mental and social well-being, and aims for all nations to reach as high on these aspects of well-being as possible (WHO 1948).

The strength of making health a human right is that all nations and territories need to focus on it. Rights as a concept in ethics has been contested, especially by ethical theories which are not deontological. From a pragmatic and constructivist point of view, human rights have been formally accepted by most nations and territories in the world (although the interpretation of them differs widely) regardless of differences in cultural or religious traditions. This has been used as an argument within ethics to justify not only the legal aspect of the declaration but also that it could be a constructivist foundation for different kinds of ethics (Nussbaum 2011).

Sustainability, one of the foundation concepts of SDGs, could be understood in two ways, either as the Brundtland definition based on anthropocentrism (discussed here) or the three-pillar model, which also includes other perspectives such as environmental ethics (further discussed in Chap. 7). The Brundtland definition has been important for the development of the UN work on sustainable development and was defined in the report *Our Common Future* (WCED 1987). Economic growth was considered the most important value that could facilitate sustainability (Purvis et al. 2019). This anthropocentric view on sustainability is echoed in Planetary Health, at least in the defining paper of Sarah Whitmee et al. (2015) and is one reason that Lerner and Berg (2017) was arguing to question whether it was a One Health approach. Although developments within the approach, Sect. 3.5 clarifies that weak anthropocentrism is still a strong foundation in Planetary health. However, the three-pillar model of sustainability connects better with the ethical imperative for One Health approaches proposed by Lederman (2018), and I will turn to that definition in Sect. 6.2.3.

4.3 Principlism

As mentioned in Chap. 3, one early suggestion of ethics within One Health approaches by me (Lerner 2013) was citing research conducted in veterinary medicine on the possibilities of sharing similar ethical frameworks in veterinary medicine and human medicine. Mullan and Main (2001) had shown that the biomedical principles proposed by Tom L. Beauchamp and James F. Childress, used in human medicine also seemed to be applicable in veterinary medicine. I should here elaborate on the original theory of principlism and analyze if it fits One Health approaches.[1]

Within bioethics and especially healthcare ethics, Beauchamp and Childress (2019) have proposed a principlism based on four principles.[2] According to the authors, the best justification method for these principles is through a reflective equilibrium deliberation (Beauchamp and Childress 2019, 439–444). The principles are respect for autonomy, nonmaleficence, beneficence, and justice.

The point with the four principles is that they recognize four aspects of bioethics which could be shared by the entire healthcare sector, regardless of a corresponding background ethical theory. Therefore, this framework has gained huge influence at least in Western healthcare and is regarded by the authors as compatible with all of the major ethical theories. This position in ethics is by some called mid-level principles (Fraser 2012), because there is no need for a foundation in an ethical theory. One can work with the principles regardless of your own personal ethical conviction if you agree on them.

The procedure for weighing the principles is case to case. The reflective equilibrium model is coined by John Rawls as a mid-level idea between theory and cases. There could be two kinds of equilibrium, a narrow and a wide, but the wide reflective equilibrium is preferred in ethics (Daniels 2016). Beauchamp and Childress describe that a wide reflective equilibrium is present when one has evaluated strengths and weaknesses in moral theories, principles, judgements, and legitimate beliefs. For each case one analyzes what is at stake so that one can specify each principle according to the case in line with a wide reflective equilibrium. None of the principles overarch the others. Each case and the specification of principles in each case decides the outcome in the weighing and which principle/s guiding the solution.

Although their work also expands and elaborates on experimental animals, human healthcare and humans are the primary focus (Beauchamp and Childress 2019, 23, 65–92). Although further developed through the years, principlism has not expanded to include either animal ethics perspectives or environmental

---

[1] Ben Mepham (2000) has developed an extension of the idea of principlism that also could be used for animals, called the Ethical Matrix, based on Beauchamp and Childress version. I do not analyze the ethical matrix in Chap. 5 because what will be evident in this Chapter is that principlism has its shortcomings when one considers One Health approaches.

[2] First version of this book came in the late 1970s. Beauchamp and Childress principlism is not the only one, several other ones have been proposed (see Veatch (2007) for a comprehensive review), but their version has been the most influential in the bioethics debate.

perspectives, even if they leave open the possibility that animals have moral status and moral value without ending in a conclusion on how and how much. I will now turn to the cases.

4.3.1 Principlism: Bats Case

In line with principlism only humans count in the Bats case. No animal ethics or environmental ethics is present.

The Principle of Respect for Autonomy
Here the conflict between western communities and local Indigenous peoples might be evident. If other communities than those living in connection with the bats disregard the autonomy of those people living there, that will be troublesome. The principle of respect for autonomy guides towards including all autonomous persons regardless of culture or religion. One could also include cultures where individual autonomy is downplayed in favor of autonomy on the group level.

The Principle of Nonmaleficence
To promote nonmaleficence one must think about what to include. The bats and the environment are not moral agents but might be seen as moral objects. The focus is on harm to humans.

The risk of catching diseases that are transferred from bats to humans is a strong driver restricting people from intervening with bats. Therefore, the main emphasis here is to avoid decisions that might harm humans indirectly if bats are harmed. Where people are dependent on bats for pollination, fertilizer etc. an eradication or disturbance of the bats will cause maleficence to humans. If habitats are destroyed in order to eradicate disease caused by bats, humans might lose livelihoods or religious places causing great maleficence.

The Principle of Beneficence
The principle of beneficence implies that one must analyze the benefits of restricting human interaction with bats, minimizing the risk of transmission of disease. This needs to be balanced with the benefits that bats contribute to humans with. These are pollination, seed dispersal, and guano production.

The Principle of Justice
Human groups living close to bats might be vulnerable as a result of intergenerational or colonial injustice. To withhold the principle of justice one must safeguard that those vulnerable groups won't suffer unfair treatment.

Weighing of the Principles
In this case, Principlism must mainly weigh between the risk of disease transmission to humans and the risk that changes in local human livelihoods (ecosystems and species distribution) will influence humans. If one can satisfy that there is no risk of violating autonomy or justice, the weighing is primarily between the aspects

of the principle of nonmaleficence and the principle of beneficence. When autonomy or justice is at risk, for example in losing livelihood for a culture or that an ecosystem will collapse due to eradication of bats, those principles will be guiding.

4.3.2 Principlism: Animal Assisted Therapy Case

The Animal Assisted Therapy (AAT) case consists of humans and animals in our care. When this case is analyzed with the framework of Principlism, animals are more or less left out due to Principlism's anthropocentric focus.

The Principle of Respect for Autonomy
AAT is used both for people with normal cognitive abilities as well as deficient cognitive abilities. This implies that not all participants in AAT will be autonomous. However, this is similar to ordinary healthcare bioethics.

The Principle of Nonmaleficence
Regarding nonmaleficence, both direct harms to humans and indirect harms arising from contact with the horses must be considered. Direct harms might be injuries from falls or kicks. Indirect harms might be feelings that occur when one witness horses suffering in their welfare.

The Principle of Beneficence
Beneficence occurs when humans get strengthened in line with the intention of the therapy. This might indirectly be facilitated by witnessing horses thriving and enjoying their life.

The Principle of Justice
Depending on the form of therapy, the principle of justice could be important. If the therapy gives rehabilitation so that one could better participate in society or avoid being a marginalized group, AAT can facilitate justice. In that case, this principle will be more guiding than the others.

Weighing of the Principles
In this case there will mainly be a weighing between the principle of nonmaleficence and beneficence. Noteworthy is that how the horses are treated might facilitate how the weighing is performed mainly due to the involved persons' subjective experiences of the treatment, but the perspective of the horse have no direct influence on the weighing.

4.3.3 Principlism: Oceans Case

In line with principlism only humans count in the Oceans case. No animal ethics or environmental ethics is present. This case has similarities with the bats case when analyzing coastal zones and islands where people live. because out on the ocean one can hardly claim that people have their livelihood. However, human actions like overfishing and pollution dumping far away from land alter or influence also the coastal zones. Overfishing might cause food shortages in places far away from the actual fishing zone. Climate change, might bring the icecap to melt resulting in overflooded coastal livelihoods.

The Principle of Respect for Autonomy
In the principle of respect for autonomy, there is a similar conflict as in the bats case. The ocean case pinpoints the conflict between economically and politically strong countries (mainly western) and vulnerable local populations. If rich nations make decisions that rules out less influential nations' autonomy, then this principle is violated.

The Principle of Nonmaleficence
All aspects, both direct and indirect, that will ruin the livelihoods of people in coastal zones are maleficent. If climate change alters the dynamics of the oceans so that sea level rises or weather becomes unpredictable, many coastal zones and islands will be flooded causing eradication of livelihoods (and even islands). With sea level rising we also get climate refugees that need to relocate their livelihoods. Other issues such as overfishing might cause people to change their culture when fishing for food needs to be replaced with other food items.

The Principle of Beneficence
The oceans are rich in resources and for example fish is regarded as an important resource for humans to thrive. Eating fish instead of ruminants would help lower the climate footprint and fish would therefore be valuable. But the beneficial aspects have another side. The oceans case shows clearly that beneficence from oceans (fishing, oil manufacturing, pollution dumping) might favor rich countries with little dependence on coastal areas or oceans, while poor countries could be worse off.

The Principle of Justice
The fact that entire island nations might be eradicated due to changes in sea level makes the principle of justice core in this case. Historical injustices in resource depletion might also influence the weighing of the principles.

Weighing of the Principles
Here, Principlism must weigh between the global south and the global north, between the livelihood of local communities and far away countries that might not suffer from the harmful effects but use the resources. The principle of justice is strongly guiding here. Decisions that now or in the near future eradicate people's livelihoods need to be carefully analyzed in order to see if people could be relocated or if that relocation is harmful or a violation to their autonomy.

4.3.4 *Conclusion Principlism*

There is a general problem in all the three cases that principlism hardly could analyze ethical aspects regarding animals, plants and ecosystems. Although I earlier suggested this framework as fruitful, there is only a limited application within the One Health approaches. For the three cases, principlism is insufficient in the Bats and Oceans cases and has only a limited fruitfulness in the AAT case. Not even in a narrow version of One Health (only incorporating human and veterinary medicine) might this be fruitful. The implication of this is that one of the most important bioethical frameworks for human healthcare, principlism, is unsuitable for One Health ethics, due to its absence of interspecies ethics.

4.4 Biophilia

As mentioned in Chap. 3, Biophilia is an ethical framework proposed for One Health approaches analyzing human relations to nature and other species (for example Maller et al. 2009; Chalmers and Dell 2015). The theory of Biophilia was developed by biologists Edward O. Wilson and Stephen R. Kellert. The core of this theory is that humans have an innate cause for valuing other species and the relations to other species.

Biophilia is more of a model of moral psychology than an ethical framework and is used today as an explanation model for why humans might reach health and well-being when being in nature (Lencastre et al. 2022) or having relations with animals (Julius et al. 2013). It also has influenced architecture by the development of biophilic design (Kellert 2018). The theory has an anthropocentric explanation of why other species and nature matters. Stephen R. Kellert summarizes this anthropocentrism as

> The self-interested basis for a human ethic of care and conservation of nature, most especially the diversity of life. (Kellert 1993a, 21)

This particular theory of biophilia was coined by biologist Edward O. Wilson in his book *Biophilia* in 1984 (Wilson 1984).[3] The theory has been further developed by Wilson (1993) as well as Kellert (1993b, 2012, 2018). In a comment on his first

[3] The term "Biophilia" was also used earlier in the work of German psychoanalyst Erich Fromm, but in a different meaning than Wilson's concept (Kellert 2012; Lencastre et al. 2022).

book on Biophilia, Wilson (1993) states that it was an attempt to link his theory of Sociobiology[4] to environmental ethics.[5] He wanted to

> …join emotion with the rational analysis of emotion in order to create a deeper and more enduring conservation ethics. (Wilson 1984, 119)

The initial definition of biophilia was

> Biophilia […is…] the innate tendency to focus on life and lifelike processes. (Wilson 1984, 1)

A further developed description is

> Biophilia, if it exists, and I believe it exists, is the innately emotional affiliation of human beings to other living organisms. Innate means hereditary and hence part of ultimate human nature. (Wilson 1993, 31)

By including Biophilia as an innate part of human nature, he argues that this psychological foundation for ethics is genetical and part of the biology of the species. In a later essay he describes these as ethics stemming from epigenetics, making ethics reduced to biological processes in humans (Wilson 2008). Although innate, Wilson (1984, 119) claims that one as a human must learn those "impulses and biased forms" that is the foundation for an environmental ethic to occur. Aiming for conservation of biodiversity (Wilson 1984), the ethical foundation is truly anthropocentric:

> I will argue the necessity of a robust and richly textured anthropocentric ethic apart from the issue of rights—one based on the hereditary needs of our own species. In addition to the well-documented utilitarian potential of wild species, the diversity of life has immense aesthetic and spiritual value. (Wilson 1993, 38–39)

Both Wilson and Kellert describe Biophilia as a need-based ethics from the human perspective broader than a utilitarian approach. The point is to preserve the humans now living and their forthcoming generations. When one does this, a conservation of our environment will follow secondarily as a result. This explains why animals, plants, species or ecosystems cannot have intrinsic values in an ethical framework based on Biophilia. The motivation for action is always based on human needs. Kellert describes this further as

> The conservation of nature is rationalized, from this perspective, not just in terms of its material and commodity benefits but, far more significantly, for the increased likelihood of

---

[4] Sociobiology was thoroughly described by Wilson in Sociobiology: The New Synthesis, which appeared 1975. Five years later, he published an abridged version to make it easier to grasp for the audience (Wilson 1980). In this book I will not go further into the critique and defence of sociobiology per se, I rather stick to discuss Biophilia and the Biophilia hypothesis. For an early attempt of gathering the positions about sociobiology, see Ruse (1979) and for a later attempt, see Driscoll (2022).

[5] Present in both Wilson's (1984) and Kellert's (1993a, b) accounts of Biophilia is the reference to Aldo Leopold's land ethic as a kind of justification,, but there seem to be no specific claim that Biophilia results in Leopold's land ethic. As will become evident in Chap. 6, Leopold's theory is ecocentric.

> fulfilling a variety of emotional, cognitive, and spiritual needs in the human animal. An ethical responsibility for conserving nature stems, therefore, from more than altruistic sympathy or compassionate concern: it is driven by a profound sense of self-interest and biological imperative. (Kellert 1993b, 60)

Because the theory explains human behavior as innate, based on self-interest and human needs no action could be altruistic.

4.4.1 Differences in Typology of Values

In the development of the Biophilia theory there has also been a development of a typology of values. Wilson (1993, quotation in Sect. 4.4) focus on three separate values—utilitarian, aesthetic and spiritual, while Kellert introduces a "typology of Biophilia values" to explain the different kinds of selfish values for others (Kellert 1993b, 59). First, Kellert (1993b) has a typology of nine values. Later, Kellert (2012, 2018) uses eight values with slightly different names and meaning. In order to get a grip on Kellert's typology of values one needs to further analyze his different versions. I have tried to summarize and compare the different versions in three of his sources, an early (Kellert 1993b), the one he preferred (Kellert 2012)[6] as well as his latest (Kellert 2018).[7] Table 4.1 visualize similarities and discrepancies in his different typologies as well as in his definitions of the values.

The values presented in Table 4.1 show a difference in whether they are positive or negative for the subject evaluating and/or the object evaluated. Starting with the 2012 version by Kellert, five values (affection, attraction, reason, symbolism, and spirituality) could be positive for both the human subject and the object (being another species or an ecosystem), one value could be negative for both the subject and the object (aversion), and two values could be positive for the subject but negative for the object (dominion, and exploitation). Aversion is maybe the most interesting value to start with because it seems like a contradiction for the theory. Aversion (earlier called negativistic value) includes feelings of fear for species and alienation from nature. This value seems to be opposed to value of attraction (formerly aesthetic value) and the earlier suggestion of the naturalistic value (Kellert 1993b).[8] This inherent tension in the typology is evident due to the rational analysis of emotions asked for, but undeveloped by Wilson. Biophilia tries to cover all versions of valuation that occur within a human being. Nevertheless, if the value of aversion is too strong that might ruin preservation of a certain species or ecosystem contra-intuitive to the aim in the Biophilia theory for nature conservation.

---

[6] As stated in Kellert (2018), the last book he wrote.

[7] Published posthumously with an afterword by his wife.

[8] Some of these values show similarities with ecosystem values, developed by the Intergovernmental Platform on Biodiversity and Ecosystem Services (IPBES). I will briefly discuss these latter values in Sect. 6.2.1.

Table 4.1 Typology of Biophilia values in Kellert (1993b, 2012, 2018). How the typologies connect to each other is based on my analysis of Kellert's texts (see Sect. 4.4.1)

Kellert (1993b)		Kellert (2012)		Kellert (2018)	
Value	Description	Value	Description	Value	Description
Humanistic	"reflects feelings of deep emotional attachment to individual elements of the natural environment" (Kellert 1993b, 52) Directed at sentient aspects	Affection	"emotional attachment, including a love of nature" (Kellert 2012, xii)	Affection	"express strong emotional attachment and at times love for features of the natural world. Commonly associated benefits include the ability to bond, care, and connect emotionally with others." (Kellert 2018, 6)
Aesthetic	"the physical beauty of nature" (Kellert 1993b, 49), "a human intuitive recognition or reaching for the ideal in nature: its harmony, symmetry, and order as a model of human experience and behavior" (Kellert 1993b, 50). Directed at sentient aspects	Attraction	"appreciation of the aesthetic appeal of nature, from a superficial sense of the pretty to a profound realization of beauty" (Kellert 2012, xii)	Attraction	"aesthetic attraction and ability to perceive beauty in nature. Associated benefits include feelings of harmony and symmetry, emotional and intellectual development, and enhanced capacities for imagination and creativity." (Kellert 2018, 6)

(continued)

Kellert (1993b)		Kellert (2012)		Kellert (2018)	
Value	Description	Value	Description	Value	Description
Ecologistic-Scientific	Ecologistic is "more integrative and less reductionistic than the scientific, involving an emphasis on interconnection and interdependence in nature as well as a related stress on integral connections between biotic and abiotic elements manifest in the flow of energy and materials within a system" (Kellert 1993b, 47), while scientific is "a greater emphasis on the physical and mechanical functioning of biophysical entities as well as a related stress on issues of morphology, taxonomy, and physiological processes" (Kellert 1993b, 48)	Reason	"the desire to know and intellectually comprehend the world, from basic facts to more complex understanding" (Kellert 2012, xii)	Intellect	"to use nature as a means for advancing rational thought and intellectual development. Benefits include cognitive skills, empirical and observational abilities, critical thinking, and learning." (Kellert 2018, 7)
Symbolic	"reflects the human use of nature as a means of facilitating communication and thought" (Kellert 1993b, 51)	Symbolism	"the symbolic representation of nature through image, language, and design" (Kellert 2012, xii)	Symbolism	"to employ the image of nature to advance communication and abstract thought. Important benefits include the capacities for language and culture, intellectual development, and enhanced imagination and creativity." (Kellert 2018, 7)
Moralistic	"encompasses strong feelings of affinity, ethical responsibility, and even reverence for the natural world … often reflects the conviction of a fundamental spiritual meaning, order, and harmony in nature" (Kellert 1993b, 53)	Spirituality	"the pursuit of meaning and purpose through connection to the world beyond our selves" (Kellert 2012, xii)	Spirituality	"to experience nature as a means for achieving a sense of meaning, purpose, and connection to creation. Associated benefits include feelings of meaningful and purposeful existence, enhanced self-confidence, and bonding with others." (Kellert 2018, 7)

(continued)

Table 4.1 (continued)

Kellert (1993b)		Kellert (2012)		Kellert (2018)	
Value	Description	Value	Description	Value	Description
Negativistic	"characterized by sentiments of fear, aversion, and antipathy toward various aspects of the natural world" (Kellert 1993b, 56)	Aversion	"antipathy toward and sometimes fearful avoidance of nature" (Kellert 2012, xii)	Aversion	"to avoid aspects of nature that generate feelings of anxiety, threat, and sometimes fear. Benefits include enhanced safety and security, coping and competitive skills, and sometimes a sense of awe- and respect for powers greater than one's own" (Kellert 2018, 6)
Dominionistic	"reflects the desire to master the natural world" (Kellert 1993b, 56)	Dominion	"the urge to master and control the natural environment" (Kellert 2012, xii)	Control	"to master, dominate, and, at times, subjugate nature. Benefits include enhanced mastery and problem-solving skills, critical thinking and cognitive development." (Kellert 2018, 7)
Utilitarian	"the physical benefits derived from nature as a fundamental basis for human sustenance, protection, and security" (Kellert 1993b, 45)	Exploitation	"the desire to utilize and materially exploit the natural world" (Kellert 2012, xii)	Exploitation	"to utilize the natural world as a source of materials and resources. Commonly associated benefits include enhanced security, extractive abilities, and practical skills." (Kellert 2018, 7)
Naturalistic	"a sense of fascination, wonder, and awe derived from an intimate experience of nature's diversity and complexity" (Kellert 1993b, 45)				

The values of dominion and exploitation are also troublesome. Both indicate the possibility of overtaxing the environment or a species, at least as it has been understood in the twentieth and twenty-first centuries. If understood like this, there is a conflict with the original idea of preserving ecosystems and species within biophilia. I leave this problem open and will return to it further down in this chapter.

If there might be these tensions between the values, one needs to have an elaborate way of considering these values in relation to each other. I will now turn to how Kellert develops these matters.

4.4.2 *Weighing of Value*

What to do with these values? Should they be combined in a certain way or should they be weighed against each other? Wilson (1984) gives no further guidance on this question, although he states that ethics will follow by acknowledging these values. Kellert (2012), on the other hand, gives some guidance. He analyses how Biophilia would turn from a description of different values into an ethical framework:

> Our consciousness reflects our values, which in turn drive our ethics. Our sense of responsibility for caring for the earth depends on recognizing how much we remain reliant on a multiplicity of connections to the natural world. An ethic that transforms our relationship to nature must embrace all our biophilic values—[all eight of them]. The core of this transformative ethic is the realization that we can achieve lives of meaning and satisfaction only by living in right relation to the world beyond ourselves. (Kellert 2012, 188)

When all eight values influence our consciousness, we will turn into ethics and take responsible care of our environment. One can see the eight values as an ethical objective list. If the full complexity is included in our consciousness, and involved in our thoughts of meaningfulness of life, the better. For this to happen, Kellert sees four conditions as necessary:

1. Engaging all our biophilic values, each revealed in balanced relation to the others, and in each adaptive and functional fashion.
2. Having a strong emotional connection to nature that reflects both a passion for and a love of life and a universe of creation.
3. Pursuing knowledge and understanding of the natural world, recognizing the limits of our intellect and the need to apply this understanding with humility and restraint.
4. Recognizing that ultimately faith and reverent relation to the natural world will be necessary for us to flourish as individuals and as a species. (Kellert 2012, 188, numbers added) [9]

I would argue that the two quotes above indicate that this framework could be seen as a form of virtue ethics. One must learn, train and master the recognition of each value. Further, one must learn how to balance the values and contemplate our mind's shortcomings by searching for a deeper meaning of all. The aspect of learning is also evident in Wilson's (1984) analysis (see Sect. 4.4). This assumption that

---

[9] Kellert also refers to Aldo Leopold arguing for a similarity between him and Leopold. As I will analyze in Chap. 6 , Leopold has an ecocentric foundation rather than Kellert's anthropocentric.

Biophilia is virtue ethics implies that Kellert's 2012 version can be interpreted as how to become wise, get the right character and flourish as individual and species with regard to conscious thinking.

Kellert's (2012) theory of balancing the values is that all values need to be present in a balanced and functional way in order for one to thrive as a human. One cannot omit any of the values out from the balancing act.[10] However, there is no golden rule for this balance. Different values could be more important depending on how different people consciously acknowledge and weigh these values. Kellert acknowledges relativity in how these values are analyzed and weighted among different people. However, not all balances are good. Only functional balances of the values are those strived for.

> All our biophilic values must be functionally revealed if we are to be healthy and whole, even if the content and priority of these adaptations will vary among individuals and groups in functional response to varying conditions, cultures, experience, biology, and ecological circumstance. Each value renders a vital contribution to the human body, mind, and spirit, and each offers a suite of instrumental advantages. When all our biophilic values occur in functional, balanced, and complementary relation to one another, an essential foundation is laid for a transformative environmental ethic that motivates us to care for nature not out of some altruistic impulse but from a profound realization of our own self-interest. (Kellert 2012, 195–196)

In this quote, all values stem from a species-egoistic and anthropocentric view. The relativity among people, partly due to genetic differences (Biophilia is the innate tendency) and partly to environmental conditions (such as culture, local settings), is fruitful for intercultural and interreligious aspects of One Health approaches. However, he avoids the question of how to settle differences. If there are more than one balance of values, which solutions should one choose? Could one solution be by appealing to the one balance based on the most developed of the four conditions? Or could there be some other process that needs to be elaborated? These are questions left unanswered.

4.4.3 Critical Remarks

In this section, I will try to address some of the theoretical critique on Biophilia before I analyze the framework with the cases. In spite of the proponents for this theory, the value of it for One Health approaches has also been contested (Kingsley and Lawson 2015 following Simaika and Samways 2010, see Sect. 3.4.1). Interestingly, those proponents for Biophilia in One Health approaches (Maller

[10] The view of balancing might at first glance seem to have some resemblance with Martha Nussbaum's Capability Approach described and analyzed in Sect. 5.5. The difference is striking, where Nussbaum talks about capabilities, Wilson and Kellert talks about a need-based ethic. Kellert assume that all values must be present and balanced, Nussbaum argues that a capability cannot be reduced down to another capability and each capability must be above a threshold for the human to thrive.

et al. 2009; Chalmers and Dell 2015; Kingsley and Lawson 2015) cite sources earlier than 2009 and have therefore not included Kellert's later and more thorough analysis. So, a call for ethics is mainly based on just the plea for ethics stemming from recognizing these values in the early version of Biophilia. Although a typology of values is present in Kellert (1993b), there are no information there on weighing balances.

As I analyze this typology of Biophilia values, I see conflicts between the values. This has earlier been suggested as one of the core concerns with the Biophilia theory (Kahn 1997). For example, if one has a strong fear of an animal species (such as aversion), that could force people to eradicate that species especially when it might be considered a disease bearer. Other individuals can ascribe several of the positive values for the object to the species and argue for preservation. Although innate, different individuals might reach different conclusions. Even more concerning, if the negativistic value is more or less innate in humans, as in the example of snakes used by Wilson (1984), all of us could reach the conclusion that all snakes should be eradicated. This would be contradictory to what Wilson aims for with his theory of Biophilia and this problem lacks a possible solution in Kellert's balancing of values because of the introduced relativity of values. One way to solve this would be to add a process of deliberating between different people to find a unified choice of actions. Another way would be to let those more trained and familiar with all their values decide using the virtue aspect of the theory.

Another critique is directed to the foundation of the theory. Holmes Rolston III (1993) questions selfish genes as the driving force for environmental ethics. He thinks Wilson's idea of innate selfishness as an explanation for protecting non-human organisms seems odd. Rolston finds that some (or many) humans tend to protect individuals very distant from themselves, i.e. no evident kinship. So, a selfish gene explanation would seem less plausible than an altruistic explanation. Following his argumentation, it became evident that selfish genes[11] cannot be extrapolated to higher levels due to several reasons, he concludes:

> We do reach a point where the quantitative expanding of self has reached a qualitative regard of a self for others with whom one is interconnected but whom one loves for what they are in themselves, not just what they are for us. (Rolston III 1993, 408)

For Rolston III, Biophilia needs an altruistic motivation to work as a foundation for environmental ethics, rather than the egoistic one proposed by Wilson. This is inherent in Rolston's view of human nature:

> Animals have the capacity to see only from their niche … [h]umans can have a transcending view from no niche. … our capacity to *see others*, to oversee a world, that distinguishes humans. (Rolston III 1993, 409, original emphasis).

From this follows his foundation for an environmental ethic expressed as:

> Skeptics and relativists may say that humans just see from another niche, and it is certainly true that when humans appraise soil and timber as resources, they see from within their

[11] Which is the foundation in Wilson's Sociobiology.

> niche. But humans also see other niches and the ecosystems that sustain the niches; they study warblers or see Earth from space. No other species has such supersight, such spectacular oversight. What humans can do that nothing else can is recognize these intrinsic values for what they are, where they are, instrumentally woven into the ecosystemic Earth. Such value, which is present, *ought* to be preserved. (Rolston III 1993, 409, original emphasis)

In my view, I agree with Rolston that the foundation of Biophilia has an inherent tension and Wilson or Kellert cannot show how the leap from selfishness to altruistic acts works. Also, reducing the field of ethics down to biological processes as Wilson (2008) suggests is not an option. The scientific endeavor has shown that reductionism could be helpful in experiments and for clearly demarcated studies, but not for more holistic studies or wicked problems (Waltner-Toews 2017). Also, the idea that Biophilia is innate has no strong evidence, rather it seems that Biophilia is learned through experience (Simaika and Samways 2010) in line with Kellert's thoughts of Biophilia.

Biophilia still might be useful in Planetary Health, due to its anthropocentric focus but hardly useful in One Health, EcoHealth and One Welfare, which are more zoo-, bio- or ecocentric.[12] As a decision framework in ethics, Biophilia lacks a structure to weigh between the different values which could be opposing and conflicting and the ethical imperative within the theory (see Kellert 2018, 189–192) lacks the possibility to develop into ethics. There needs to be something more than just a typology of values to be useful as an ethical framework. Let us turn to the cases.

4.4.4 *Biophilia: Bats Case*

The Bats case consists of humans and bats in their environment. When this case is analyzed with the framework of Biophilia, human values towards bats and their environment is analyzed. The typology of values that Kellert himself favoured (Kellert 2012, Table 4.1) is used to analyze the Bats case.

Affection
The value of affection could occur when one is fascinated over the bats' abilities to fly and echolocate in the environment. One could also feel affection for the caves where the bats live.

Attraction
The value of attraction may either be present or absent. Bats might be seen as aesthetic animals because of their swiftness in flight, but they could also be labeled as unattractive and ugly. Some refer to bats as "flying rats", where rats in this case often is an animal negatively valued. In this case, the value of fear from diseases (see the value of symbolism below) and the popular culture link between bats and blood sucking might influence people more, making attraction harder to get.

[12] See Chap. 2 for the arguments why these approaches could be seen to have this value foundation.

Reason
In those cases where humans know that bats are important parts of their ecosystem, this knowledge could result in the value of reason.

Symbolism
The present problem in the One Health approaches is that the virus hunters place a negative value of symbolism on bats as those that might be the most prone to transfer disease. An example is the children's book *Living safely with bats* (Francisco et al. 2018).

Spirituality
In Indigenous cultures, the caves where bats hibernate could be religious places. In their holistic views of nature, these places need to be preserved intact in order to preserve the value of spirituality, i.e. the religious and spiritual aspects of their culture and life.

Aversion
As mentioned in the analysis of the symbolism value and the affection value, there are several reasons why bats may cause the value of aversion to people. If these reasons are exaggerated, the total analysis of the values might be skewed. This will result in a strong value of aversion and unfairly diminishing the values of symbolism and affection.

Dominion
The dominion value is about control. There might be a wish to control the ranges of bat populations, especially concerning infectious zoonoses that might cause pandemics. This might influence where bats occur with the result that their ability to pollinate plants will be diminished.

Exploitation
Several aspects could give value of exploitation. Bats contribute to humans as providing fertilizers for agriculture. Guano gathering from bats is an important source to facilitate crops in regions where they live. Many species of bats feed on fruits and contribute therefore with important pollination and seed dispersal for human needs.

Conclusion
As I analyze the Biophilia values, I see conflicts between the values. In the bats case, if one has a strong fear of bats (aversion), one won't reach any positive value (for example spirituality or attraction). That could force people to eradicate a bat species especially when it also might be considered as a disease bearer, which inflict on human self-interest. Other individuals can ascribe both value of symbolism and reason to the bat species and argue for preservation.

4.4.5 Biophilia: Animal Assisted Therapy Case

The Animal Assisted Therapy (AAT) case consists of humans and animals in our care. When this case is analyzed with the framework of Biophilia, human values towards the animal involved in the therapy are analyzed. The typology of values that Kellert himself favored (Kellert 2012, Table 4.1) is used in this analysis.

Affection
Emotional attachment and love for an animal is one of core aspects in animal assisted therapy. Some therapies are based on these premises and one goal of AAT could be facilitating the value of affection to increase the ability to feel empathy.

Attraction
For the value of attraction, horses could be experienced as aesthetically beautiful. A riding session when the human and the horse cooperate seamlessly, where no harm occurs on either, could be a pleasure to look at.

Reason
Animal assisted therapy might favor the value of reason. For example, this value has been a goal when training people who have suffered from depression, fatigue syndrome, and stroke.

Symbolism
Horses have been so important in culture that they today have a symbolic value for humans and have changed parts of human history (Walker 2008).

Spirituality
In this case, the value of spirituality could be understood as meaningfulness in life. Horses might be the meaning of life for several humans. They might be the motivator for people experiencing fatigue who need a reason to get a grip on themselves. Going to the stables might be the reason to leave a bed to do an activity during the day. Another aspect is linked to reducing stress and experiencing mindfulness. Riding activities might facilitate the possibility to contemplate on life and what is good life, as well as meeting one self's inner soul.

Aversion
People might have a value of aversion of horses due to the risks involved (being bitten, kicked, or falling of a saddle). This might hamper the therapy.

Dominion
Riding as part of animal assisted therapy might be seen as a form of dominion. It is not natural but could be justified in anthropocentric ethics. In almost all equine sports, a view of human dominance is present. In animal assisted therapy on the other hand there is a more dynamic view where one can use the idea that sometimes the human lead and sometimes the horse. People participating in this therapy might have either a strong dominion view or a weak dominion view on the value of the animal.

Exploitation
Horses will be used to fulfil human needs in healthcare, which may result in an exploitation value. If horses are instrumentalized, humans might use them as they wish, which is something that should be prohibited in AAT (IAHAIO 2018).

Conclusion
Compared to the Bats case, the AAT case has no particular tensions between the values. There is a possibility to design an AAT so that it is ethically sound.

4.4.6 Biophilia: Oceans Case

The Oceans case consists of humans and their relation to oceans. When this case is analyzed with the framework of Biophilia, human values towards aspects of the oceans are analyzed. The typology of values that Kellert himself favored (Kellert 2012, Table 4.1) is used to analyze the Oceans case.

Affection
The value of affection could result in a deep emotional attachment to the sea. This has been conveyed in lyrics, literature, and art. A theme could be that the human is connected to the sea. The big blue with no land in sight might cause a sense of fascination and awe for something that is bigger than us and that we cannot live in permanently. This could also facilitate the values of attraction as well as reason.

Attraction
Oceans might have an aesthetic value when one observes the extended water towards the horizon. Scuba divers may get value of attraction when experiencing the beautiful colored world below water with all its life-forms. This has also been used both in art and in literature.

Reason
For the value of reason, the sea seems to inspire us to reflect. Trying to grasp what it is to be a human. This indicates that the value of reason could be improved in these matters if one uses oceans as a case for One Health deliberations.

Symbolism
The ocean is a symbol for where all life originated which resulted in humans. Oceans can also be seen as lungs for the world and a healthy ocean might save us from climate change. However, traditionally, oceans and their ecosystems and resources have been regarded as inexhaustible, always being able to bounce back. Garbage dumping, polluting, and overtaxing the populations of the sea has therefore been far more performed than in terrestrial environments. We are fully aware of these risks, although it seems that humans have less understanding and will to mitigate these problems.

Spirituality
The ocean and the animals therein are in some cultures part of their religion and give daily meaning to their lives, which is part of their value of spirituality.

Aversion
Feelings of aversion for oceans could be linked to the risk of flooding or drowning. Some may fear that species could become extinct, or that ecosystems will be lost, ruining livelihoods.

Dominion
In the dominion value there is for example the urge to tame the elements; to outwit the big catch or kill the whale. Whale hunting has been an issue internationally for several years. This must be weighed against the risk of depleting animal species. In sports fishing, they are trying to develop catch- and release-methods with as little harm as possible.

Exploitation
If the Oceans are as important as the rainforests, the exploitation value should be to preserve the ecosystems within, for the future of humankind. Also, the risk of the rising of the temperatures causing flooding might backfire on humans and therefore carefulness with emissions of greenhouse gases must be continued.

Many cultures are dependent on ocean resources for their daily life. One must also remember that many people use oceans for recreation, either swimming, fishing or cruzing. Turning oceans into fish farms would also risk the flora and fauna beneath and might limit ocean resources in the long run.

Conclusion
In this case there might appear several conflicts between and within values. Depending on how one experiences oceans, as an opportunity or a threat, that influences how to balance the values. If, for example, the problem of overfishing is guiding in the evaluation, this habit is explained as the dominion value being too strong in the process of weighing.

4.4.7 Conclusion Biophilia

There is a problem with this framework. It is easy to develop a list of values, but how to reach an ethical decision is unclear. Due to the lack of guidance in how to weigh the different values this framework is hard to use. Also, the foundation on egoistic values might be highly questionable for One Health approaches. Maybe it may work with Planetary Health, but I would dismiss this framework for One Health approaches. In my opinion, the ethics of Biophilia tend to be some kind of virtue ethics although not fully developed into a functionable ethical framework.

4.5 Summary

In this chapter of analyzing anthropocentric frameworks, I have dismissed principlism as too narrow as a fruitful ethic for One Health approaches. The main reason is that it only applies to clinical settings in human medicine and veterinary medicine (although it may work in a narrow conception of One Health). For One Health approaches, one needs a more inclusive ethic with another foundation. Biophilia might be useful within Planetary Health, but the framework lacks the possibility to be an ethical framework for decision making. I have therefore found that both Principlism and Biophilia are unsuitable as an ethical foundation for One Health approaches. They are non-inclusive and violate the claim made by Lederman (2018) that One Health approaches need an ethic that value humans, animals and ecosystems.

However, human rights need to be further considered based on the pragmatic argument that the United Nations considers health as a human right. This implies that a weak anthropocentrism must at least be considered. Also, if Biophilia's psychological aspect holds, humans are interested in relationships with animals and the environment and that also could be an argument not to rule out weak anthropocentrism already. Strong anthropocentrism is not possible because it conflicts with the ethical imperative of One Health approaches which states that humans, animals and ecosystems should be valued. Next Chapter will deal with zoocentric frameworks.

References

Beauchamp, Tom L., and James F. Childress. 2019. *Principles of biomedical ethics*. 8th ed. Oxford: Oxford University Press.

Chalmers, Darlene, and Colleen Anne Dell. 2015. Applying one health to the study of animal-assisted interventions. *EcoHealth* 12:560–562.

Daniels, Norman. 2016. Reflective equilibrium. In *Stanford encyclopedia of philosophy*. Stanford, CA: Stanford University.

Driscoll, Catherine. 2022. Sociobiology. In *Stanford encyclopedia of philosophy*. Stanford, CA: Stanford University.

Francisco, Leilani, Ava Sullivan, Jonathan Goley, and Stephanie Martinez. 2018. Living safely with bats. https://ohi.vetmed.ucdavis.edu/programs-projects/predict-project/publications. Retrieved 10 Oct 2022.

Fraser, David. 2012. A "practical" ethic for animals. *Journal of Agricultural and Environmental Ethics* 25:721–746.

International Association of Human-Animal Interaction Organizations (IAHAIO). 2018. The IAHAIO definitions for animal assisted intervention and guideline for wellness of animals involved in AAI. IAHAIO White Paper 2014, updated 2018. https://iahaio.org/wp/wp-content/uploads/2021/01/iahaio-white-paper-2018-english.pdf. Retrieved 15 Aug 2022.

Julius, Henri, Andrea Beetz, Kurt Kotrschal, Dennis Turner, and Kerstin Uvnäs-Moberg. 2013. *Attachment to pets: An integrative view of human-animal relationships with implications for therapeutic practice*. Cambridge, MA: Hogrefe.

Kahn, Peter H, Jr.. 1997. Developmental psychology and the biophilia hypothesis: Children's affiliation with nature. *Developmental Review* 17:1–61.

Kellert, Stephen R. 1993a. Introduction. In *The biophilia hypothesis*, ed. Stephen R. Kellert and Edward O. Wilson, 20–27. Washington, DC: Island Press.

Kellert, Stephen R. 1993b. The biological basis for human values of nature. In *The biophilia hypothesis*, ed. Stephen R. Kellert and Edward O. Wilson, 42–69. Washington, DC: Island Press.

Kellert, Stephen R. 2012. *Birthright: People and nature in the modern world.* New Haven, CT: Yale University Press.

Kellert, Stephen R. 2018. *Nature by design: The practice of biophilic design.* New Haven, CT: Yale University Press.

Kingsley, Jonathan, and Justin Lawson. 2015. Finding a unified understanding of nature. *EcoHealth* 12:551–552.

Lederman, Zohar. 2018. Stamping out culling: The ethics of culling from a one health perspective. Ph. D. Thesis. Centre for Biomedical Ethics, National University of Singapore.

Lencastre, Marina P. A., Diogo G. Vidal, Rui Estrada, Nelson Barros, Rui L. Maia, and Paulo Farinha-Marques. 2022. The biophilia hypothesis explored: Regenerative urban green spaces and well-being in a Portuguese sample. *International Journal of Environmental Studies.* https://doi.org/10.1080/00207233.2022.2067411.

Lerner, Henrik. 2013. The philosophical roots of the "One Medicine" movement: An analysis of some relevant ideas by Rudolf Virchow and Calvin Schwabe with their modern implications. *Studia Philosophica Estonica* 6 (2): 97–109.

Lerner, Henrik, and Charlotte Berg. 2017. A comparison of three holistic approaches to health: One Health, EcoHealth, and Planetary Health. *Frontiers in Veterinary Science* 4:163. https://doi.org/10.3389/fvets.2017.00163.

Maller, Cecily J., Claire Henderson-Wilson, and Mardie Townsend. 2009. Rediscovering nature in everyday settings: Or how to create healthy environments and healthy people. *EcoHealth* 6:553–556.

Mepham, Ben. 2000. A framework for the ethical analysis of novel foods: The ethical matrix. *Journal of Agricultural and Environmental Ethics* 12:165–176.

Mullan, S., and D. Main. 2001. Principles of ethical decision-making in veterinary practice. *In Practice* 23:394–401.

Nussbaum, Martha C. 2011. *Creating capabilities: The human development approach.* Cambridge, MA: The Belknap Press of Harvard University Press.

Purvis, Ben, Yong Mao, and Darren Robinson. 2019. Three pillars of sustainability: In search of conceptual origins. *Sustainability Science* 14:681–695.

Rolston, Holmes, III. 1993. Biophilia, selfish genes, shared values. In *The biophilia hypothesis*, ed. Stephen R. Kellert and Edward O. Wilson, 381–414. Washington, DC: Island Press.

Ruse, Michael. 1979. *Sociobiology: Sense or Nonsense?* Dordrecht: D. Reidel.

Simaika, John P., and Michael J. Samways. 2010. Biophilia as a universal ethic for conserving biodiversity. *Conservation Biology* 24(3):903–906.

United Nations (UN). 1948. Universal Declaration of Human Rights. www.un.org/en/about-us/universal-declaration-of-human-rights. retrieved 12 Sept 2023.

United Nations (UN). 2023. Make the SDGs a reality. https://sdgs.un.org. Retrieved 5 Dec 2023.

Veatch, Robert M. 2007. How many principles for bioethics? In *Principles of Health Care Ethics*, ed. Richard E. Ashcroft, Angus Dawson, Heather Draper, and John R. McMillan, 43–50. Chichester: Wiley.

Walker, Elaine. 2008. *Horse.* London: Reaktion Books.

Waltner-Toews, David. 2017. Zoonoses, One Health and complexity: Wicked problems and constructive conflict. *Philosophical Transactions of the Royal Society B* 372:20160171.

Whitmee, Sarah, Andy Haines, Chris Beyrer, Frederick Boltz, Anthony G. Capon, Braulio Ferreira de Souza Dias, Alex Ezeh, Howard Frumkin, Peng Gong, Peter Head, Richard Horton, Georgina M. Mace, Robert Marten, Samuel S. Myers, Sania Nishtar, Steven A. Osofsky, Subhrendu K. Pattanayak, Montira J. Pongsiri, Cristina Romanelli, Agnes Soucat, Jeanette Vega, and Derek Yach. 2015. Safeguarding human health in the Anthropocene epoch: Report of The Rockefeller Foundation-Lancet Commission on Planetary Health. *Lancet* 386:1973–2028.

Wilson, Edward O. 1980. *Sociobiology: The abridged version*. Cambridge, MA: Belknap Press of Harvard University Press.
Wilson, Edward O. 1984. *Biophilia*. Cambridge, MA: Harvard University Press.
Wilson, Edward O. 1993. Biophilia and the conservation ethic. In *The biophilia hypothesis*, ed. Stephen R. Kellert and Edward O. Wilson, 31–41. Washington, DC: Island Press.
Wilson, Edward O. 2008. The nature of human nature. In *Biophilic design: The theory, science, and practice of bringing buildings to life*, ed. Stephen R. Kellert, Judith H. Heerwagen, and Martin L. Mador, 21-25. Hoboken, NJ: Wiley.
World Commission on Environment and Development (WCED). 1987. *Our common future*. Oxford, UK: Oxford University Press.
WHO. 1948. *Official Records of the World Health Organization, no. 2*, 100. Geneva: WHO.

Chapter 5
Animal Ethics, Veterinarian Ethics and Humans

> *Interest in human-animal relations is growing across contexts and cultures. Certain people and communities have always recognized that humans are but one of many species that share this planet. Others are now more fully recognizing this fact and reflecting on what it means for individuals, whole species, and our shared futures. (Coulter 2016, 1)*

Abstract Here, zoocentric ethical frameworks will be discussed and their value for One Health will be analyzed. Ethical frameworks concerning animals have originated mostly from two fields, animal ethics, and the science of animal health and welfare. One could differ between two kinds of views on animals. One is on species similarities where those proponents focus on similarities between humans and other species. The other is on species differences, where proponents argue that one must consider species-specificness to avoid lumping all animals in the same bowl and acknowledging specific traits that might be as important as human traits. As an example of one of the first philosophers following this view, I briefly present Bernard Rollin, which gained influence in the scientific discussions of animal welfare. For One Health approaches, the latter position is more accurate and won't hinder ascribing moral status to different species. This chapter's in-depth analysis will be on Martha Nussbaum's Capability Approach and David Fraser's Four principles of animal ethics, which both have been proposed as fruitful within the One Health approaches.

Keywords Capability approach · Martha Nussbaum · Principlism · David Fraser · Veterinary ethics · Bernard Rollin · Zoocentrism

H. Lerner, *Ethics for One Health Approaches: A Roadmap for Future Directions*, Advancing Global Bioethics 22,
https://doi.org/10.1007/978-3-032-05239-1_5

5.1 Introduction

The last chapter was the first of the normative analysis of frameworks proposed within One Health approaches. In this Chapter, I continue this task by turning to zoocentric ethical frameworks within the wide area of animal ethics (including veterinarian ethics). The expansion of the moral community to also include animals and give them moral status are, as Kendra Coulter states in the epigraph, a history of two strands. Some authors claim that there has during times been a profound interest in the welfare and value of animals (Thomas 1984), while others claim that interest in animals has grown since the 1970s (Singer 2011).

The procedure for this Chapter is similar to Chap. 4. After an overview, each of the proposed frameworks will be presented in its original form and then developed or criticized depending on content or suitability to One Health approaches, and finally applied to the three cases which were introduced in Sect. 1.4. This final application will indicate strengths and weaknesses of the framework analyzed.

5.2 Overview of Zoocentric Ethical Frameworks

Ethical issues about animals have been discussed both in the scientific field of animal health and welfare[1] and in ethics.[2] In the scientific field of animal health and welfare emphasis has been on the concept of animal welfare which has several connotations, but all are rather holistic focusing on the animal and the surroundings (Lerner 2008). Within the scope of animal welfare, physical, mental, social and environmental issues are discussed. There is also a specific branch of ethics, veterinary ethics, focusing on professional matters for veterinarians, relations between clients, veterinarians and the animal involved, as well as the view on animals and how it is valued (Rollin 1999).

Kirsten Schmidt (2011) has argued that the scientific discussion and theory formation on the concept of animal welfare has had limited influence in the core animal ethics discussion due to several theoretical matters based on differences in research approach, use of word (welfare has a very specific connotation for a certain kind of ethical theories in ethics, where in animal welfare it refers to several kind of theories) and on a difference in how to value facts and arguments. In her analysis Schmidt establish the inextricable connection between lines of argumentation in animal welfare compared to ethics. In this Chapter, I will discuss both the ethical

[1] For a full description of this see Lerner (2008, 15). A shorter version 'animal welfare science' is often used, but that term neglects the importance of the concept of health within the field.

[2] I here focus on two of the main strands of research on ethical aspects of human-animal relations. In one study analyzing general research on human-animal relationships the authors found 27 distinctive scientific fields (Echeverri et al. 2018). This diverse arena must in the future be more bridged (if possible) if one wants a unified view of ethics, but for the aim of this book I settle with my simplified distinction.

framework originating from animal ethics as well as the scientific field of animal health and welfare. Main emphasis has been on animals living in human care and less emphasis on wild animals.[3]

For animals, there have been two strands of ethical argumentation during the twentieth and twenty-first century. These could be separated by analyzing the foundation of ethical consideration, the question on why animals are included within the moral sphere. One strand is arguing that similarities between animals and humans make them valuable (for example Tom Regan (1983), Richard Ryder (1998) and Peter Singer (2011)).[4] The other strand states that the differences between species makes them valuable (for example Bernard Rollin (1995), Donna Haraway (2003), Jacques Derrida (2008), Henrik Lerner (2008), and Martha Nussbaum (2022)).[5] The first one extrapolates specific traits that are important for humans and includes animals that share these. The second one acknowledges that species have evolved differently and specific traits that humans do not have could be as important as human ones. In this line, otherness as well as relations could be important concepts to link to ethics. Here, not sharing some important traits might not be an exclusion from moral status, such as the case in anthropocentric ethics and zoocentric ethics based on similarities. Rather, it might be a pathway for respect and consideration in ethical dilemmas.

First, I will briefly argue why the first line, species similarities, might be less suited for One Health approaches and try to give the reason why I dismiss these frameworks in favor of frameworks based on species differences.

5.3 Species Similarities

There are many ethical frameworks that build their theory that animals have moral status on the fact that humans and at least some animals share similar features. Two of these similar features have been *suffering* and *rationality*. Suffering as a common denominator for humans and animals originated in utilitarianism and have

---

[3] Although this is changing. For examples on discussing ethical issues for wild animals from both the science of animal health and welfare as well as from animal ethics, see Berg et al. (2020) and Nussbaum (2022).

[4] Peter Singer together with co-workers (Coghlan et al. 2021) has also joined the One Health ethics debate. However, despite arguing for a wider moral circle together with some suggestions to what to include, this paper is just a call for more ethics and not a proposal of an ethical framework. This was also briefly discussed in Chap. 3.

[5] Note that the Biophilia theory analyzed in Chap. 4 differs from this distinction, because it defines that humans has the tendency to focus on other species inherent. This inherent tendency is used as a foundation why we should have a concern for other species.

proponents within the animal welfare science.[6] Rationality originated in deontological ethics claiming rights for animals if they have a sufficient level of rationality.[7]

To use a similarity-approach has pedagogical value.

An argument that begins by things humans value and demonstrates that some animals share these traits, ultimately leading to the conclusion that if we value these qualities in humans, we should also value them in animals, is a compelling way to gain support. However, it is misleading and could be counterargued in several instances.

By valuing human traits, one miss to value species-specific traits that are not shared by humans. There is a risk of anthropomorphism when focusing only on similarities.[8] Only a limited number of species could be included, even if one stretches and make the concept of suffering or rationality more rudimentary. Still, not all animals might be included despite the extrapolation top-down from humans. Also, the concept might be too diluted to be interesting for humans.

To meet the claim made in Chap. 3 on including multispecies justice within One Health ethics there seem to be a demand for a view on these matters as species differences. Donna Haraway (2018) argues elsewhere that an inclusive ethics needs to focus also on those that are different from oneself, and this also holds for One Health ethics.

For One Health approaches, which need an ethic that encompass animal species from the entire animal kingdom (considering ecological processes between species) and also need to solve cases where there might be conflict between animals that could harm us (for example as in the Bats case), an ethic that favours only human traits would be less fruitful than an ethic focusing on animal (and human) differences.

Therefore, I turn to promising frameworks that focus on species differences. I will start with Bernard Rollin, because his focus on species-specificness has been influential. However, his framework is not applicable to One Health approaches. I will instead discuss David Fraser's and Martha Nussbaum's ethical frameworks, which better suit the One Health approaches. Rollin is a combination of deontologist and virtue ethicist while Fraser is more of pragmatist suggesting fruitful principles (compare this to Beauchamp and Childress in Chap. 4) and Nussbaum, finally, is a virtue ethicist.

---

[6] Some of the proponents of this view are Peter Singer, Richard Ryder, and Marian Stamp Dawkins.

[7] Some of the proponents of this view are Tom Regan, and Christine Korsgaard.

[8] One can argue that there is no specific humanness. That is a valid argument, we share all our "specialities" with at least some species at a rudimentary level (such as toolmaking, funerals etc.). Still, as humans we have a specific neural system and a specific range of our senses which gives us a specific way of understanding ourselves. Other species have other versions of neural systems and ranges of senses. A bee, a bird or a cat experiences the world differently.

The biologist Jacob von Uexküll (2001) proposed the theory of an animal's *Umwelt* which encompass this. This has been partly accepted within animal welfare science and is also in line with phenomenological strands of thought. I have earlier elaborated on *Umwelt* to stress the importance to not treat animals as a uniform group neither in the discussion of health and welfare in animal nor in animal ethics (Lerner 2008, 19–28).

5.4 Species Differences

One of the first to lift the problem with lumping all animal species in the same bowl was philosopher, physicist, and biophysicist Bernard E. Rollin.[9] In his book *The Unheeded Cry. Animal Consciousness, Animal Pain and Science* (Rollin 1989), he introduces the idea that animal species have certain needs based on their nature.

> We would expect its behaviour to be appropriate to its *telos*, the unique, evolutionary determined, genetically encoded, environmentally shaped set of needs and interests which characterize the animal in question—the 'pigness' of the pig, the 'dogness' of the dog, and so on. (Rollin 1989, 146, original emphasis)

The use of *telos* indicates his closeness to Aristotle, although I have argued at length elsewhere, that his concept differs from the Aristotelean concept (Lerner 2008). However, the crucial part for this book is his introduction of the terms species specific or natural behavior[10] which if thwarted could cause suffering because of needs being unmet (Rollin 2005). Needs should therefore not be understood only in physical terms, rather he emphasizes the importance of meeting mental needs (Rollin 1989). In his framework, one ethical issue was the treatment of domestic animals.

> On my view, one can in principle assess the genetic similarity of the wild to the domestic. If they are close, yet the living conditions are significantly different for the domestic animal, then one may have a prima-facie reason to believe that the animals' *telos* is being violated—that a square peg is being forced into a round hole—and that it is not living as it evolved to do. (Rollin 1989, 257, original emphasis)

Here, it becomes evidently clear that the limitations of a species depending on their species-specific abilities are important to understand. The phrase already quoted above of "the pigness of a pig, the dogness of a dog" is used in his later writings (for example Rollin 2000) together with the warning of not putting square pegs in round holes to visualize the problem of thwarting an animals species-specific or natural behavior.

If the telos is unique for each animal species, then humans would also have a specific telos. Rollin states that there is a difference to his view that "it is probably easier to grasp the nature of a dog or a pig than to encapsulate 'human nature'" (Rollin 1999, 43).

> Animal telos is of course not the same as human telos; thus the protections they require are not the same, thus the rights of animals cannot be the same as the rights of humans. But the fact that animals do have interests that are as important to them as speech and belief are to

[9]This section on Bernard Rollin is a condensed and rewritten part of Chapter 8 in my doctoral thesis (Lerner 2008, 141–163), which was an in-depth critical investigation of his concept of animal welfare and his view on animal nature. That analysis draws me to the conclusions presented in this section which differs slightly from some other author's narrower presentations of his work focusing only on the "natural" aspect in his theory.

[10]The concepts of the nature of animals and natural behavior has had different connotations within veterinary medicine and could also be linked to the discussion in human medicine on the concept of normality and norms (Lerner and Hofmann 2011).

> us is indubitable. Social animals need to be with others of their kind; animals built to run need to run; these interests are species specific. Others are ubiquitous in all species with brains and nervous systems—the interest in avoiding pain, in food and water, and so forth. (Rollin 1995, 159)

Rollin thinks that humans have unique interests such as speech and belief that differentiate them from other animals. Also, he claims that some interests in animals may be as important as speech and belief in humans (Rollin 1995). This resembles the view that I (Lerner 2008) favor about the possibility of unique preferences in certain animals. Further, each telos is species specific—dogness of a dog, pigness of a pig etc.—and thus normatively states that each individual of a species needs to be analysed with regard to its own species (making it similar to the claims by Martha Nussbaum developed in Sect. 5.5.1). Thus, one may possibly criticise Rollin for being a speciesist, but the framework in its ideal form respects each animal's nature with a focus on that individual animal.

However, Rollin has mainly been acknowledged as important within the science of animal health and welfare and gained much less reputation in animal ethics. Now I turn to discuss the two suggested frameworks by Martha Nussbaum and David Fraser, which are built on species differences.

5.5 Capability Approach

Martha Nussbaum's *capability approach* was regarded as suitable for One Welfare (Sect. 3.6). The capability approach originates from the idea that all humans have the right to dignity in their lives and every human has combined capabilities that help that individual to reach a dignified life. She argues that her theory at least is compatible with western and Indian traditions of thought (Nussbaum 2011) in order to set the possibility to reach compliance in the world.

Each combined capability consists of internal capabilities (such as personality, intellectual and physical abilities, perception, knowledge etc.) and surrounding conditions (such as social, political, and economic circumstances), and requirements (such as infrastructure, legislation, and corresponding social norms) (Nussbaum 2011).

According to the framework, a reduction of a combined capability below the acceptable threshold will hinder the individual having dignity and one cannot compensate for the lack of one capability with excess of another. Dignity is present when all ten combined capabilities of her objective list[11] is at least above an acceptable threshold level.[12] The capabilities listed are:

---

[11] Cf. Sect. 4.4.2 where I in footnote 10 compare Nussbaum's objective list of capabilities with the theory of Biophilia to indicate the differences between these two.

[12] Dignity could be defined with different connotations and there has been attempts to understand and distinguish between these. Philosopher Lennart Nordenfelt (2021) distinguishes between four different notions in the philosophical tradition, (1) Dignity of merit, (2) Dignity of moral stature,

1. Capability of life
2. Capability of bodily health
3. Capability of bodily integrity
4. Capability of senses, imagination and thought
5. Capability of emotions
6. Capability of practical reason
7. Capability of affiliation
8. Capability of other species
9. Capability of play
10. Capability of control over one's environment (Nussbaum 2004, 2011, 2022)

Although Nussbaum does not group the capabilities, I think it might be fruitful to see three areas of concern among the capabilities if one thinks of health in a wide sense. Capability 1–3 mainly refers to bodily health (although there are mental and social aspects relevant), capability 4–6 refers to mental health, and capability 7–10 refers to social aspects of health. I will specify these capabilities in Sect. 5.5.2, but before that I need the next section to present and consider her expansion of the capability approach to other species than humans.

5.5.1 Which Other Species Are Covered?

Although originally developed for humans, Nussbaum (2004, 2011, 2022) clearly states that the combined capabilities could also be applied to animals. For animals, it is necessary to analyze each species' requirements for the different combined capabilities as well as the acceptable threshold levels. She clearly states that flourishing animals with their capabilities above the threshold also have dignity (Nussbaum 2022). Because dignity follows from a situation where the individual has all her capabilities above the specific thresholds and each species should have their own version of the proposed capabilities, I believe we could here talk about species-specific dignity.[13]

Compare this to Bernard Rollin, mentioned in Sect. 5.4. When he refers to dogness of a dog and cowness of a cow, he only refers to a whole, the *telos* of the species with its species-specific or natural behaviour. In Nussbaum's framework, species-specific or natural behaviour is distinguished in ten different capabilities.

---

(3) Dignity of identity, and (4) Human dignity. In Nordenfelt's analysis of Nussbaum's theory he argues that her version belongs to two of the notions, namely human dignity and dignity of identity.

[13] Compared to the distinction made by Nordenfelt, one could argue that dignity of identity will only apply in the animal case due to human dignity being a species-specific concept only applicable to humans. However, following Nussbaum one can argue that each species has its own version of dignity based on what it is to be that species, and each kind of species-specific dignity is valued. The idea of human dignity could then be understood as one version of species-specific dignity. This could be developed also within other ethical frameworks based on species differences, for example by using Bernard Rollin's theory.

This is a much finer and specified grid which could analyze a case sufficiently. The capabilities are the same for humans and animals able to have capabilities, the differences is the interpretation of each of the capabilities. Each capability needs to be in line with each particular species species-specific or natural behaviour or abilities.

An important note needs to be made here. Although Nussbaum talks about species, only individual animals count and have moral status in her ethical framework. This is the same as for humans. The human species is only considered when deciding the interpretation on each capability. In ethical deliberations, humans are considered on an individual level. Dignity and value are therefore attributed to individuals.[14]

Not all animals are able to flourish. Animals able to flourish have at least significant striving, where this term is defined as

> be[ing] capable of perception and desire, and of moving in response to that combination (Nussbaum 2022, 119)

This definition follows an Aristotelian view of animals (when she also cites a passage from *On the movement of Animals*). Nussbaum further defines perception as

> the capacity to focus on objects in the world, in a way that's not just a causal collision, that has real directedness [...] intentionality. The world *looks like something* to these creatures. They have some sort of subjective experience. (Nussbaum 2022, 119, original emphasis)

Desire is defined as

> the creatures we are looking for don't just mechanically jump away from harm or move toward food; they have a felt orientation toward what is seen as good and a felt aversion to what is seen as bad (Nussbaum 2022, 119)

Besides humans, Nussbaum (2022) seems to tentatively regard at least mammals, birds, fish, reptiles, cephalopods and insects (at least bees) to have moral status. This is based on current biological knowledge, namely neuroanatomy and behavior studies. She is unsure about crustaceans and rule out cnidarians and porifera. Plants are excluded, she cannot give them moral status, although she proposes that they could be given ethical concern similar to the protection of species or the way we could value an ecosystem. In the next section, there will be an analysis of each capability for humans (Nussbaum 2011, 2022), and animals in general (Nussbaum 2004, 2022). In the cases, species-specific analyses will follow.

In my own work (Lerner 2019), analyzing human-animal interactions within equine assisted interventions, Nussbaum's capability approach was developed into a comprehensive human-animal approach of evaluation. When one wants to decide if an animal assisted intervention is ethically sound one can go through each of the combined capabilities for all involved animals and all involved humans to decide if the capability is *reduced*, *preserved*, or *improved* depending on the interpretation of each species capability list.

---

[14] This is similar to how Bernard Rollin treats the species level. It is the thwarting of the individual's telos (not the species) that counts as something wrong to do. Value is attributed to individuals and not to the species.

Considering all capabilities, one is then able to decide. If at least one of them in one of the individuals (regardless of species) is reduced to below the threshold level for that capability, the intervention is harmful and not ethically sound. Ethically non-harmful occurs when all capabilities are at least preserved without any of the capabilities for any of the individual being reduced below the threshold level. Ethically beneficial is when at least one of them in one of the individuals are improved without any of the capabilities for any individual being reduced below the threshold level.

5.5.2 Capabilities Explained for Humans and Animals

In this section, I will present interpretations for each of the capabilities for both humans and animals (in a more general sense, an in-depth analysis for horses are elaborated in the AAT case in Sect. 5.5.5).[15]

Life concerns being able to live a life with normal species-specific length without dying before life has been so reduced that it is not worth living (Nussbaum 2004, 2011). Life is a capability that all animals can share regardless of whether they are sentient or not. One could also argue that this capability is a prerequisite, without it no other capability will be there (Lerner and Silfverberg 2019).

Bodily health implies being able to have good health. One needs proper food and a sufficient place to live. In animals, Nussbaum (2004) puts emphasis on avoiding cruel and ill treatment of animals, i.e. common animal welfare aspects.

Bodily integrity is the ability to move freely from place to place without restrictions. In the restrictions, Nussbaum includes violations of the body, and sexual or violent abuse. All these should be avoided. In animals freedom from abuse is core.

Senses, imagination, and thought, is the ability to use one's senses, to think and reason influenced by the senses. Therefore, freedom of speech and artistic expression is central for this capability. Freedom of choices are important for animals.

Emotion is being capable of loving others. Included in this capability is also the ability to grieve a loss of a dear friend. In animals, emotions such as joy, envy, and gratitude can be found.

Practical reason is the ability to form and reflect on the idea of the good life and how it is planned. That is one of the capabilities that can be hard to find in animals.

Affiliation has two aspects in humans. One is to live with and towards others, the other is self-respect and freedom from humiliation (Nussbaum 2004). Bonds between individuals of the same species is the corollary in animals.

Other species is the capability to have concern for other species. This is present in both humans and animals.

---

[15] This compilation is rewritten from part of the text in Lerner and Silfverberg (2019).

Play is about being able to enjoy recreational activities and be able to play. Nussbaum share the view that both humans and sentient animals are able to play.

Control over one's environment is for humans divided in two aspects. The first is the ability to participate in politics, while the second is being able to own property. For animals, the first aspect concerns respecting the animal in itself. This could be protected by animal rights. The second aspect is that the animals should be safeguarded from having their home environment destroyed.

5.5.3 Capabilities for Other than Humans: Some Development and Critique

Her adaptation of her framework to animals might be discussed and somewhat questioned. First, I will discuss the problem inherent in her early texts where she seemed to reduce the capabilities into each other. Second, I will discuss whether the different ranges of senses in animals might need any addition to her framework.

Lerner and Silfverberg's (2019) analysis of Nussbaum's earlier texts (Nussbaum 2004, 2011) indicated an inherent problem that when she proposed examples for animals, she used similar examples on several capabilities. This could be understood as reducing the number of capabilities, which is contrary to her claim that all capabilities must be present.

However, using the horse as an example species, Lerner and Silfverberg were able to establish that all capabilities had a possibility to be ascribed to horses, even though research on cognition in horses have been less conducted than in other domestic animals. Those capabilities that might be troublesome for horses are *practical reason*, *affinity*, and *control over one's environment*. What I can see is that it depends on how one rephrases these for the horse. If one avoids an anthropocentric intention of these concepts, they might be easier to attribute to animals. Nussbaum seems to hold this position in her latest book on these matters (Nussbaum 2022). But this is only held for sentient animals.

To extend the capabilities approach to cover the entire animal kingdom, one needs to be clear that several of the capabilities becomes diluted and might disappear when one moves from sentient animals to non-sentient animals. The solution for Nussbaum (2022) is therefore to restrict her framework to animals able to flourish and this is the reason why she draws a line in the animal kingdom as described in Sect. 5.5.1.

Nussbaum holds that the list is open and not fixed. There could be additions if needed. However, at the same moment she uses the list as fixed, saying that no one yet has come up with a better solution. Therefore, her list is unchanged through her texts (Nussbaum 2004, 2011, 2022). So, if that is the case for humans for which this framework has mainly been discussed, what will be the case when the framework is extended to animals?

What happens to the fact that other sentient animals have another range of senses so that they experience the world differently than we do? There are senses present in the animal kingdom such as sensing magnetic fields or electrical potentials that we lack (Keeley 2002; Lerner 2008). Animals also occupy other niches in the environment and have other abilities linked to their constitution of the body. I have argued elsewhere (Lerner 2008) that at least differences in senses might influence animal ethics and that it might be anthropocentric to exclude these aspects for example when one think about what should be counted in ethics. Would this imply that some capabilities might be species-specific?

Nussbaum opens for this but seems to stick to the idea that those ten capabilities are so general that they will cover also all aspects of animals. In her early developments (Nussbaum 2004, 2011) she tries to rephrase the description of each capability for the human case in interpretations suitable for the animal case. Animals are treated as a group without further specification. This has been contested by Lerner and Silfverberg (2019), who argue that one must analyze each capability for a specific species because. Species differences are far more important than interspecies similarities. In her recent book on the matter (Nussbaum 2022), she argues that her framework is focusing on species differences and that each species could have their own interpretation of the capability list. Species differences are therefore why animals have moral status, not species similarities. Still, capabilities are tied to the individual level (as well as moral status) and not to the species level. Her way of analyzing the differences is through a limited number of cases used through the book, trying to make thick descriptions of ethical matters for particular individuals, which even are named with human names. She also states that thick ethological descriptions of individuals of an animal species are key in understanding the species itself. This is quite similar to the attempt from Lerner and Silfverberg (2019) to try to find the rephrased capabilities for one species, the horse, and use the capability approach to consider interspecies ethics between humans and horses. However, additions to the list of capabilities have not yet been proposed.

If additions to the list might be suitable, are retractions from the list possible? An animal that is non-sentient could for example have a shorter list of capabilities that is needed for a flourishing life excluding all capabilities based on sentience. If the list of capabilities is brought forward from analyzing a certain species, that species are the benchmark for setting the list. For humans, this has resulted in ten capabilities. If we avoid anthropocentrism and have the species as benchmark, we would arrange a list for each species. Depending on senses and abilities for that species, each species could have differences both in the interpretations of capabilities in the capability list as well as the number of capabilities at least in theory.[16] As stated above, Nussbaum seem to avoid that conclusion by arguing for animals that flourish. However, flourish is defined in anthropocentric terms, with humans as benchmark. This might be a troublesome concern for her framework.

---

[16] See also footnote 8 in this Chapter.

5.5.4 *Capability Approach: Bats Case*

Here follows the analysis of the Bats case when Nussbaum's framework of capabilities (Sect. 5.5) is applied together with the evaluation method described by me (Sect. 5.5.1). In line with the evaluation method, one must deliberate issues that risk reducing a capability below a threshold (including also losing a capability).

Capability of Life
Human capability of life might be preserved or improved if avoidance of bats minimizes risk of disease transmission. Killing of bats might influence pollination of important food sources for humans which might reduce this capability for humans.

Bat capability of life will be reduced, if human fear for disease transmission results in culling bats to minimize the risks.

Capability of Bodily Health
The worst risk for the capability of bodily health is the risk of disease transmission. Bats are hosts for several diseases and regarded as the origin source for Covid-19.

Bats require suitable breeding areas, an adequate food supply, and ability to rest and hibernate to preserve their bodily health. There are also indications that bats might change behaviour to avoid transmitting disease to others in the roost (Moreno et al. 2021).

Capability of Bodily Integrity
For those that regard caves where bats rest as sacred places, a restriction to visit will reduce the capability of bodily integrity as well as the capability of senses, imagination and thought.

To preserve the capability of bodily integrity in bats, one should avoid chasing them away from their caves or eating places even though one think that this could minimize disease transmission.

Capability of Senses, Imagination and Thought
Eradicating bats might reduce the capability of senses, imagination and thought in humans. The possibility of seeing bats could improve imagination and thought when one thinks about how it is to be a bat, which is living a life different than ours, flying at night dependent on echolocation.

For bats, dependent on their audial environment, humans might disturb their sense of hearing and change their soundscape making it more difficult to find food.

Capability of Emotions
If bats are strongly associated with being seen as carrier of diseases, fear of bats might be shared by many. This would be especially troublesome in areas where there is close contact between humans and bats.

Capability of Practical Reason
The capability of practical reason is not expected to be influenced for humans within this case.

One example of the practical reason for bats might be the possibility to have a "sick leave" to avoid transmitting disease (Moreno et al. 2021).

Capability of Affiliation

Some cultures see the habitats of bats as sacred areas. These aspects need to be considered and protected in order to preserve and improve the capability of affiliation. This is also tightly linked to freedom of religion within human rights.

Destroying colonies or making colonies disperse reduces bats' capability of affiliation.

Capability of Other Species

In some cultures, close ties between humans and bats exist. A fear of bats or even worse, a ban on interactions with bats will reduce this capability of other species.

Bats have most probably relations to other species although bats lacks mutual relationships with humans.

Capability of Play

For the capability of play, no direct influence on humans is expected.

With the destruction of habitats, bats might lose this capability.

Capability of Control over One's Environment

Indigenous peoples who live in the vicinity of bats might feel restrictions in their livelihoods resulting in a reduced capability of control over one's environment. This capability might also be affected by the fact that they are a vulnerable group and the marginalization tendency in western societies for this group might be stronger and they might lose more political control.

With the destruction of habitats, bats might lose this capability.

5.5.5 Capability Approach: Animal Assisted Therapy Case

Here follows the analysis of the Animal Assisted Therapy case when Nussbaum's framework of capabilities (Sect. 5.5) is applied together with the evaluation method described by me (Sect. 5.5.1). In line with the evaluation method, one must deliberate issues that risk reducing a capability below a threshold (including also losing a capability).[17]

Capability of Life

Capability of life could be improved by the physical training in AAT sessions. This occurs specifically in horseback riding, because that activity trains the entire body at the same time. However, being with horses and riding might reduce the capability of life if one avoids following safety rules. In an ethically sound AAT, accidents that risk shorten either the life of the human or the horse should always be avoided.

[17] This compilation is rewritten from Lerner and Silfverberg (2019). For further references see the original text.

Horses' length of life might be reduced by having too long and too many sessions in AAT.

Capability of Bodily Health

One of the main goals in AAT with horses is to improve the capability of bodily health for humans. This could be done either while working with the horses in the stable or when riding.

Proper diet and treatment are important to preserve bodily health in horses. Overweight in humans might cause suffering in horses in riding sessions. This could be avoided by using practices for humans without riding or choosing a proper horse according to the weight of the client. Pain in horses is often a factor behind accidents when the rider is thrown off. Knowledge of signs of exhaustion and pain, along with the ability to allow a horse to rest when it shows these signs, is important for preserving or improving this capability.

Capability of Bodily Integrity

The capability of bodily integrity can be improved in cases when a person gets courage and will to perform actions that has been lost due to the AAT.

For horses, one must observe the horse well not to force the horse do things it wants to avoid.

Capability of Senses, Imagination and Thought

The capability of senses, imagination, and thought is often improved in horseback riding because riding is a complex activity demanding the rider's persistence, concentration, and focus.

Horses are today seen as having a rich cognitive life. Giving the horse the ability to explore and solve problems is an important aspect of AAT which might improve the horse's capability.

Capability of Emotions

The capability of emotions can be improved if the icebreaker effect occurs. This is when a person that cannot talk about emotions suddenly open up and starts talking. This effect could be facilitated by AAT. Emotional attachment to the horse is often also a part of AAT, and this is especially important for humans in vulnerable situations.

Horses seem to have a rich emotional life and are also susceptible to human emotions. If they experience prolonged negative emotions from humans, this capability would be reduced.

Capability of Practical Reason

In AAT, this capability of practical reason could be improved by letting the human train to choose activities and find motivating aspects that later could improve quality of life.

This is one of the capabilities which is tough to ascribe to horses. One explanation is that there is less cognitive research in horses than in other domestic species. However, horses seem to have dynamic leadership in flocks which might indicate the ability to have a practical reason.

Capability of Affiliation

The capability of affiliation in AAT is here seen as the possibility of being together with other humans in a shared activity. This could be improved by the setting of the AAT.

Horses need affiliation with other horses, and solitary horses have a high risk of developing stereotypies.

Capability of Other Species

The capability of other species is a prerequisite for both human and horse in AAT and is improved in the therapy.

Capability of Play

AAT might improve the capability to play for humans depending on how one chooses activities.

Play behavior in horses should however be avoided in most sessions, because it can cause damages in humans. Better is to give the horses leisure time between sessions where they could play to preserve this capability.

Capability of Control over One's Environment

The capability of control over one's environment for humans is improved and trained in all therapies with adequate method for rehabilitation and treatment.

In horses, this capability could be reduced if humans hinder the horse's exploration behavior. Then the horse might be less interested in showing intentional behavior.

5.5.6 Capability Approach: Oceans Case

Here follows the analysis of the Oceans case when Nussbaum's framework of capabilities (Sect. 5.5) is applied together with the evaluation method described by me (Sect. 5.5.1). In line with the evaluation method, one must deliberate issues that risk reducing a capability below a threshold (including also losing a capability). In the Oceans case one must choose those animals that could be said to have capabilities. Within the diverse animal groups in the ocean, Nussbaum has placed particular emphasis on whales (Nussbaum 2022).

Capability of Life

Changes in climate that results in sea level rising can destroy livelihoods by flooding that might reduce human capability of life. Extreme weather due to climate change might cause storms and big waves which can crush people and their livelihoods.

Flourishing animals risk their life and populations by being fished or killed. Pollution, such as oil leakages, might kill seabirds.

Capability of Bodily Health
A healthy ocean is a prerequisite for people being able to live on the food resources in the ocean. Pollution, such as microplastics, might reduce the capability of bodily health both in humans and in animals.

Among the flourishing animals, dolphins has been affected by fishing methods causing reduction in bodily health and bodily integrity with the risk of losing life. Efforts are being made to find better fishing methods that protects dolphins.

Capability of Bodily Integrity
Oil pollution in the sea influences seabirds and sea mammals by causing harm to their bodies. Cleaning of tanks at sea should be avoided as much as possible.

Capability of Senses, Imagination and Thought
This capability might not directly be influenced, but if the ocean that for some people are experienced as untouched and pristine gets too polluted with less species this might be the case. Imagination and thought might be hampered due to this.

Capability of Emotions
Killing of whales, seeing dead fish and polluted (oil, plastics) sea causes feelings of aversion in many humans.

Capability of Practical Reason
The capability of practical reason might not be inflicted in humans. This capability is also the hardest to apply on animals, mainly due to the difficulties to get this knowledge about other species.

Capability of Affiliation
For whales which rely on audio communication in their flocks, sound pollution in the water by boats influences their social relationships and they might risk losing each other's positions, which reduce the capability of affiliation.

Capability of Other Species
Due to the difference of habitat and the fact that marine animals cannot live on land, and we cannot live in sea, the capability of other species seem to be less important to analyze. However, human affection to dolphins is an important aspect to think about. The rich flora of stories about interactions between dolphins and humans indicate that both species might actively seek each other (see for example Nunny and Simmonds 2019). If dolphins get extinct this will reduce this capability.

Capability of Play
Capability of play might not be influenced at all for humans. For animals in sea, changes might influence animals' possibility to express play behavior. There are also examples of invitations from wild dolphins to humans for interspecies play (Nunny and Simmonds 2019) that might get reduced.

Capability of Control over One's Environment
Climate change with sea level rise and extreme weather reduces the capability of control over one's environment.

For sentient animals, climate change with heatwaves and sea level rising changes their environment, which results in less control over it reducing this capability.

5.5.7 *Conclusion Capability Approach*

The limit of this ethical framework is that only humans and flourishing animals are truly covered. Most of the capabilities refer to cognitive aspects of life and presupposes a nervous system. Therefore, an extension to plants and ecosystems is impossible. So, as evident in the cases, Nussbaum's capability approach might only work with sentient animals and humans. This make it more troublesome to analyse in the Oceans case because it is more vague which animals that should be considered. However, this might be a fruitful ethics to add in order to safeguard aspects that relate to individuals with flourishing lives to avoid reductionism, for example in human-animal relations. This could be important in One Welfare.

5.6 Four Principles of Animal Ethics

David Fraser has been influential as a theorist in animal health and welfare science where he proposed a comprehensive view on the concept of animal welfare. He has also suggested a principles-based ethics for animals.[18] Based on mid-level principles, similar to the principlism discussed in Sect. 4.3, he suggests a pluralistic ethical framework of four principles for our relations to animals (Fraser 2012). He sees it as a "practical approach" with no ethical theory foundation needed. His principlism results from how people view their relations to animals and what we know about these relations. As a starting point for his ethical discussion, he recognizes four main areas of concern in how humans affect animals, which are

1. keeping animals, for example, on farms and as companions,
2. causing intentional harm to animals, for example through slaughter and hunting,
3. causing direct but unintended harm to animals, for example by cropping practices and vehicle collisions, and
4. harming animals indirectly by disturbing life-sustaining processes and balances of nature, for example by habitat destruction and climate change (Fraser 2012, 721)

In these four areas, the treatment of animals and the issues are different from each other. Therefore, there are differences in kinds of ethical concerns raised within these areas which implies that ethical guidelines should rather recognize these differences. For Fraser, this results in four principles, each of them in line with each way of affecting animals above. One has

[18] For an introduction to animal welfare and the influence of David Fraser on this discussion, see Lerner (2008), Schmidt (2011), and Fraser (2023).

1. to provide good lives for the animals in our care,
2. to treat suffering with compassion,
3. to be mindful of unseen harm, and
4. to protect the life-sustaining processes and balances of nature. (Fraser 2012, 721)

All these principles are written in an open manner allowing several possible interpretations. The third and fourth principle are further elaborated in a more recent paper (Fraser 2019), providing deeper insight into which kinds of ethical concerns these principles could handle and give possible practical solutions. He also argued that an ethic able to balance intended versus unintended harms as well as individual versus collective actions in the field of animals and ecosystems might narrow the difference between zoocentric and ecocentric frameworks (Fraser 2019, 7). However, making his principles more precise or specific is not an ambition in that text.

Fraser has in his first text specified two aspects that are guiding for the principles.

> …(other things being equal) long-lasting effects are more significant than brief effects, and … irreversible harms should be deemed more serious than reversible harms (Fraser 2012, 739)

In his most recent text, he extends the ideas on how to use the principles:

> …making ethical decisions about animals would not be based on any single principle but rather would involve deciding, in a reflective, consultative and science-informed manner, how to balance the various principles in complex real-world situations (Fraser 2023, 312)

Fraser holds that a kind of weighing is performed, and that more than one principle are considered in real-world cases ruling out the interpretation of his framework that each principle refers to a specific area of concern (domestic animals, wild animals or environment). There is also a strong emphasis that the ethical valuation should be based on scientific knowledge and empirical information (the science-informed manner). However, Fraser avoids clearly demonstrating how the principles could be applied in specific cases. Instead, we must turn to Anne Fawcett et al. (2018) for such a development.

5.6.1 Development and Application of the Four Principles

Anne Fawcett et al. has further developed the framework of Fraser's four principles when they analyzed three ethical cases for veterinary practice (Fawcett et al. 2018). Framing Fraser's ethic within the One Welfare approach they argue that the four principles incorporate already working ethical theories—i.e. utilitarianism, deontology, relational ethics, ethics of care (feminist ethics), virtue ethics and some aspects of environmental ethics (especially conservation issues) (Fawcett et al. 2018,

6)—which makes them more useful in real world problems.[19] Still, they argue a core consequentialist focus:

> Fraser's "practical ethic" requires that we consider which approach is likely to yield the least total harm. To this end, it employs a consequentialist approach that requires the weighing-up of harms and benefits. (Fawcett et al. 2018, 4)

In Fawcett et al. version one weighs the principles in each case (similar to the principlism proposed by Beauchamp and Childress, see Sect. 4.3). Fawcett et al. (2018) compared Fraser's four principles with the Ethical Matrix proposed by Ben Mepham as well as the precautionary principle.[20] Compared to those, the core strengths in Fraser's four principles are its focus on unintended or unseen harm and incorporating human well-being, animal welfare, and environmental sustainability. If this is the case, they claim that one can bridge animal ethics with at least the conservation issues within environmental ethics. The reason is that the four principles are not entirely individualistic, which is explained in the following quote:

> This is why Fraser's "practical" ethic requires us to act at the level of the individual animal (to provide good lives for animals in our care and to treat suffering with compassion, to be mindful of unseen harm (to individuals)), but also more broadly to be mindful of unseen harm (to others) and to protect the life-sustaining processes and the balances of nature. (Fawcett et al. 2018, 11)

They also provided a pedagogical schematic way of analyzing cases. To has a sufficient grasp on knowledge and value aspects as well as different perspectives on the matter, they suggested to make a description (called "specific considerations") for each of the principles and put them above each other in a table. This tabulation makes it easier to follow which information that refer to each principle. By comparing the different rows (one for each principle) one gets a richer description of the case. The richer the better when weighing (Fawcett et al. 2018).

---

[19] These three real world examples are animal hoarding, farmer neglecting stock, and treatment and release of wildlife. In García Pinillos (2018) description of ethical areas of concern within One Welfare, the first case belong to the area "Social implications of improved animal welfare", the second case might belong to the third area "Animal health and welfare, human well-being, food security and sustainability", and the third case to the fifth area "Sustainability: connections between biodiversity, the environment, animal welfare and human well-being" (see Sect. 3.6).

[20] The Ethical Matrix is an extension of Beauchamp and Childress principlism to cover animal ethics (Mepham 2000). It is based on three principles, the principle of wellbeing, the principle of autonomy, and the principle of justice. The first principle is a combination of Beauchamp and Childress principles of beneficence and nonmaleficence.

The precautionary principle is used when there are uncertainties about the risks with specific actions. If an action is assumed to cause for example environmental harms or risks then that act might be omitted from the decision.

5.6.2 Critical Remarks

When one analyzes the four principles in both Fraser's version and the version by Fawcett et al. one must consider that they are mid-level principles that do not need an ethical theory foundation. One must adopt a constructivist position, and if there is agreement on the principles to be used, the ethical background of those deliberating them can vary. It might even be a strength that those deliberating them have different ethical backgrounds so that one can grasp a wider variety of perspectives and obstacles in the case making the description holistic.

The main critique is that the method requires a consequentialist approach to weigh aspects drawn from multiple ethical theories, which reduces these ethical perspectives to a simple weighing of harm. This weighing might bias certain aspects from those perspectives so that they will be given insufficient weight. Therefore, the concept of harm needs to be further developed in this theory to encompass non-consequentialist values, such as value of rights, dignity, or virtue, in order to fulfill the claim of Fawcett et al that the four principles can incorporate all ethical theories.

5.6.3 Four Principles of Animal Ethics: Bats Case

In the Bats case only three of the four principles are applicable. The principle to provide good lives for animals in our care is inapplicable. In Table 5.1 a tabulation is done in line with the table method proposed by Fawcett et al. (2018) for Fraser's four principles.

5.6.4 Four Principles of Animal Ethics: Animal Assisted Therapy Case

In the Animal Assisted Therapy case all four principles are applicable, although the principle to protect the life-sustaining processes and balances of nature might have less importance. In Table 5.2 a tabulation is done in line with the table method proposed by Fawcett et al. (2018) for Fraser's four principles.

5.6.5 Four Principles of Animal Ethics: Oceans Case

In the Oceans case only three of the four principles are applicable. The principle to provide good lives for animals in our care is inapplicable. In Table 5.3 a tabulation is done in line with the table method proposed by Fawcett et al. (2018) for Fraser's four principles.

Table 5.1 Bats case analysed with the method of applying Fraser's four principles proposed by Fawcett et al. (2018)

Principle	Specific considerations
Provide good lives for animals in our care	Because bats are wild animals, this principle will not apply
Treat suffering with compassion	Both human and animal suffering must be acknowledged and avoided as much as possible. For humans, the suffering caused by transmitting disease must be compared to the suffering that will occur if people don't get access to the things they gain from bats. For the bats, the suffering of an altered environment or eradication of individuals to minimize the risk of transmission of disease must be considered
Be mindful of unseen harm	Here the risk that bats as a whole group of species gets negatively value laden as bearers of disease needs to be considered. A de-valuation of a species might in the long run result in unseen harm Bats which are eco-locating and dependent on their audial environment might suffer from changes in the soundscape of the landscape. Because the human sense of hearing and the bat sense of hearing differ, we might have difficulties to understand the harm we as humans do
Protect the life-sustaining processes and balances of nature	Here one must gather information on how bats contribute to life-sustaining processes and the balance of the ecosystem to avoid irreversible effects. Bats are important pollinators, contribute with nutrients and are important in the ecosystem balance

Table 5.2 Animal assisted therapy (AAT) case analysed with the method of applying Fraser's four principles proposed by Fawcett et al. (2018)

Principle	Specific considerations
Provide good lives for animals in our care	In AAT one must provide as good welfare for the animal and as good well-being for the human being as possible. This includes physical, mental, and social aspects in best available practice
Treat suffering with compassion	When horses work in AAT it is important to value them as co-worker to avoid instrumentalization and the risk of seeing them as tools. If the animal is seen as co-worker, one has more compassion to the animal, and it is easier to not over-work the animal. One should never work with horses in pain, they should rest
Be mindful of unseen harm	Unexpected happenings must be prevented. All things that could scare the horse should be considered because a scared horse could cause damage to either itself or any of the humans participating in the AAT
Protect the life-sustaining processes and balances of nature	This principle might have less importance. For indoor activities one must analyse whether the activity cause environmental hazards. For outdoor activities one has to think about the risk of overgrazing or trampling which will cause the plant cover to disappear risking soil erosion

Table 5.3 Oceans case analysed with the method of applying Fraser's four principles proposed by Fawcett et al. (2018)

Principle	Specific considerations
Provide good lives for animals in our care	No animals in the oceans are in our care, therefore this principle will not apply
Treat suffering with compassion	Animals that suffer in the ocean must be acknowledged and treated with compassion. Fishing methods must cause as little suffering as possible
Be mindful of unseen harm	This principle might be harder to interpret below water than above. Historically there seems to have been a phycological tendency that risks of harming has been downplayed in oceans and a misconception that the sea always copes and bounces back. Examples of this could be attitudes such as; "You cannot overfish a population" or "if we put waste in deep waters it won't hurt". Nowadays we know that most fish populations are heavily disturbed by fishing and waste is a core problem in the oceans. Be mindful of unseen harm needs one to think more bravely in an environment where we land-based humans have less experience One has to think about pollution such as waste (including microplastics) and sounds. Waste might trap fish or birds. Animals might swallow waste. Sound pollution might disturb how whales communicate and find each other in the oceans Heat waves might strike animals as bad in water as on land Climate change might alter currents changing food availability for certain fish species
Protect the life-sustaining processes and balances of nature	Overfishing for human and animal food risk the balance between species in the oceans Oil pollution might kill animals directly influencing food webs in the ecosystem

5.6.6 Conclusion Four Principles of Animal Ethics

Both the bats and the oceans case demonstrate the problem with Fraser's specific first principle, the focus on animals in human care. For wild animals one deals primarily with the other three principles. Only in the AAT case one deals with all four principles. This indicates that this theory best suits human-animal interactions with animals in human care. Also, in this framework, there is a one-sidedness. Focus is primarily on the animal and humans are omitted.

So, a general conclusion on principlism follows from the analysis in Chap. 4 as well as the analysis in this Chapter. All forms of principlism tend to be too narrow in scope and not inclusive enough for One Health approaches.[21] However, there could still be some cases in narrow One Health or One Welfare where this framework could be useful.

[21] This also holds for the popular framework in animal ethics called the Ethical Matrix proposed by Ben Mepham (2000). Because his theory is an extended version including animals of the Beauchamp and Childress four principles, the critique emphasized on their principlism (regarding the scope of the principle and values included) are also applicable for the Ethical Matrix.

5.7 Summary

In this Chapter, ethical zoocentric frameworks have been analyzed. The findings can be summarized as follows. For One Health approaches, a focus on species differences is far better than species similarities.

The Capability Approach, although troublesome if one wants to extend the moral community to plants, could be applicable in at least One Welfare and fruitful for human-animal relationships at least regarding sentient animals. The strength is that it is built on species differences at least in the interpretation of the capabilities for different species. One Health approaches also might need an ethical approach which recognize the rich diversity in human life (and in other animal species).

David Fraser's four principles of animal ethics are insufficient for One Health approaches. Like other versions of principlism, it is too narrow to fit the claims for a One Health ethical framework. Next Chapter will deal with environmental ethics in biocentric or ecocentric frameworks.

References

Berg, Charlotte, Henrik Lerner, Andrew Butterworth, and Chris Walzer. 2020. Editorial: Wildlife welfare. *Frontiers in Veterinary Science* 7:576095. https://doi.org/10.3389/fvets.2020.576095.

Coghlan, Simon, Benjamin John Coghlan, Anthony Capon, and Peter Singer. 2021. A bolder one health: Expanding the moral circle to optimize health for all. *One Health Outlook* 3:21.

Coulter, Kendra. 2016. *Animals, work, and the promise of interspecies solidarity*. Basingstoke, UK: Palgrave Macmillan.

Derrida, Jacques. 2008. *The animal that therefore i am*. New York: Fordham University Press.

Echeverri, Alejandra, Daniel S. Karp, Robin Naidoo, Jiaying Zhao, and Kai M. A. Chan. 2018. Approaching human-animal relationships from multiple angles: A synthetic perspective. *Biological Conservation* 224:50–62.

Fawcett, Anne, Siobhan Mullan, and Paul McGreevy. 2018. Application of Fraser's "practical" ethic in veterinary practice, and its compatibility with a "One Welfare" framework. *Animals* 8:109.

Fraser, David. 2012. A "practical" ethic for animals. *Journal of Agricultural and Environmental Ethics* 25:721–746.

Fraser, David. 2019. Why we need a new ethic for animals. *Journal of Applied Animal Ethics Research* 1:7–21.

Fraser, David. 2023. *Understanding animal welfare: The science in its cultural context*. 2nd ed. Chichester: Wiley.

García Pinillos, Rebeca. 2018. *One welfare: A framework to improve animal welfare and human well-being*. Wallingford: CABI.

Haraway, Donna. 2003. *The companion species manifesto: Dogs, people, and significant otherness*. Chicago, IL: Prickly Paradigm Press.

Haraway, Donna. 2018. Staying with the trouble for multispecies environmental justice. *Dialogues in Human Geography* 8 (1): 102–105.

Keeley, Brian L. 2002. Making sense of the senses: Individuating modalities in humans and other animals. *The Journal of Philosophy* 99 (1): 5–28.

Lerner, Henrik. 2008. The concepts of health, well-being and welfare as applied to animals. A philosophical analysis of the concepts with regard to the differences between animals.

Ph.D. Thesis. Linköping: Department of Medical and Health Sciences, Division of Health and Society, Linköping University. Linköping Studies in Arts and Science No. 438.

Lerner, Henrik. 2019. A proposal for a comprehensive human–animal approach of evaluation for animal-assisted interventions. *International Journal of Environmental Research and Public Health* 16:4305. https://doi.org/10.3390/ijerph16224305.

Lerner, Henrik, and Bjørn Hofmann. 2011. Normality and naturalness: A comparison of the meanings of concepts used within veterinary medicine and human medicine. *Theoretical Medicine and Bioethics* 32:403–412.

Lerner, Henrik, and Gunilla Silfverberg. 2019. Martha Nussbaum's capability approach and equine assisted therapy: An analysis for both human and the horse. In *Equine cultures in transition: Ethical questions*, ed. J. Bornemark, P. Andersson, and U. von Ekström Essen, 57–68. London: Routledge.

Mepham, Ben. 2000. A framework for the ethical analysis of novel foods: The ethical matrix. *Journal of Agricultural and Environmental Ethics* 12:165–176.

Moreno, Kelsey R., Maya Weinberg, Lee Harten, Valeria B. Salinas, L. Gerardo Ramos, M. Herrera, Gábor Á. Czirják, and Yossi Yovel. 2021. Sick bats stay home alone: Fruit bats practice social distancing when faced with an immunological challenge. *Annals of the New York Academy of Sciences* 1505:178–190.

Nordenfelt, Lennart. 2021. The concepts of dignity: An analysis. Ersta Sköndal högskola arbetsrapportserie nr 99.

Nunny, Laetitia, and Mark Simmonds. 2019. A global reassessment of solitary-sociable dolphins. *Frontiers in Veterinary Science* 5:331.

Nussbaum, Martha C. 2004. Beyond "compassion and humanity": Justice for non-human animals. In *Animal rights: Current debates and new directions*, ed. Cass R. Sunstein and Martha C. Nussbaum, 299–320. Oxford: Oxford University Press.

Nussbaum, Martha C. 2011. *Creating capabilities: The human development approach*. Cambridge, MA: The Belknap Press of Harvard University Press.

Nussbaum, Martha C. 2022. *Justice for animals: Our collective responsibility*. New York: Simon & Schuster.

Regan, Tom. 1983. *The case for animal rights*. Berkeley, CA: University of California Press.

Rollin, Bernard E. 1989. *The unheeded cry. Animal consciousness, animal pain and science*. Oxford: Oxford University Press.

Rollin, Bernard E. 1995. *The frankenstein syndrome: Ethical and social issues in the genetic engineering of animals*. Cambridge: Cambridge University Press.

Rollin, Bernard E. 1999. *An introduction to veterinary medical ethics: Theory and cases*. Ames, IA: Iowa State University Press.

Rollin, Bernard E. 2000. The ethics of pain control in companion animals. In *Animal pain. A practice-oriented approach to an effective pain control in animals*, ed. L. J. Hellebrekers, 17–38. Utrecht: van der Wees.

Rollin, Bernard E. 2005. Animal happiness: A philosophical view. In *Mental health and well-being in animals*, ed. Franklin D. McMillan, 235–241. Ames, IA: Blackwell.

Ryder, Richard D. 1998. *The political animal: The conquest of speciesism*. Jefferson, NC: McFarland.

Schmidt, Kirsten. 2011. Concepts of animal welfare in relation to positions in animal ethics. *Acta Biotheoretica* 59 (2): 153–171.

Singer, Peter. 2011. *Practical Ethics*. 3rd ed. Cambridge: Cambridge University Press.

Thomas, Keith. 1984. *Man and the natural world: Changing attitudes in england 1500-1800*, Penguin Books edition. London: Penguin.

von Uexküll, Jacob. 2001. An introduction to Umwelt. *Semiotica* 134 (1/4): 107–110.

Chapter 6
Environmental Ethics

The human being has to handle a sensitive, finely tuned, living oneness. The health of the human being connects to the earth's; she cannot disconnect that connection however she carries herself. (Wägner 2007[1941], 356). (My translation from Swedish. Original quote "Människan har att handskas med en känslig, finstämd levande enhet. Människans hälsa går tillbaka till jordens, hon kan inte göra sig kvitt det sammanhanget hur hon än bär sig åt.")

Abstract In this chapter I introduce environmental ethics and make a brief overview of the field. In environmental ethics, not only individuals count, but also other levels such as ecosystems and species might count. Discussions on values are common, and I will go through the Intergovernmental Platform on Biodiversity and Ecosystem Services (IPBES) analysis of instrumental, intrinsic, and relational values. Then, an in-depth analysis of Aldo Leopold's Land Ethic follows, which ends in a proposal of a Land Ethic Decision Model which I apply on the three ethical cases that are used in the book.

Keywords Relational values · Ecocentrism · Biocentrism · Sustainability · Nature's contribution to peoples · Nature futures framework · Aldo Leopold · Land ethic

H. Lerner, *Ethics for One Health Approaches: A Roadmap for Future Directions*, Advancing Global Bioethics 22,
https://doi.org/10.1007/978-3-032-05239-1_6

6.1 Introduction

After analyzing zoocentric frameworks, I now turn to biocentric or ecocentric frameworks belonging to environmental ethics. As in the two previous chapters, the procedure is similar. I will start with an overview of environmental ethics. Then, each proposed framework will be presented in its original form and then developed or criticized due to its relevance or suitability for the One Health approaches. Finally, each of the frameworks will be applied to the three cases presented in Sect. 1.4. This final application will indicate strengths and weaknesses of the framework analyzed.

6.2 Overview of Environmental Ethics

Environmental ethics is sometimes traced back to the American natural conservationist Aldo Leopold who in the first half of the 1900s developed his ethics on ecosystems. Swedish Elin Wägner, one of the first environmental ethicists and ecofeminists and contemporary to Leopold, connects in this chapter's epigraph to the oneness of the world.

If anthropocentric ethics mainly concerns humans, one species among mammals, and zoocentric ethics concerns many animals (or only some groups of animals), the rest of the organisms existing on earth will be left to environmental ethics. Although the number of kingdoms (where Animalia is one kingdom) are not settled, one probable division of organisms is into seven kingdoms (Ruggiero et al. 2015). Environmental ethics is holistic and consider ethical aspects related to all seven kingdoms (Archea, Bacteria, Chromista, Fungi, Plantae, and Animalia).

In environmental ethics, one differs between biocentrism and ecocentrism. Biocentrism values living organisms. Moral status has those that have a life. One of the earliest biocentrists was theologian and philosopher Albert Schweitzer (1875–1957). Ecocentric frameworks have less focus on the individual level and more focus on the population and systemic level. Ecosystems as well as species are within the primary discussion and individuals are therefore downplayed. One important feature with biocentric and ecocentric frameworks is that value could be ascribed to other entities than individuals (contrary to most anthropocentric and zoocentric frameworks). Environmental ethicist Holmes Rolston III, who has worked with a typology of values within environmental ethics, states the possibility to ascribe value on different levels, such as the animal, organism, species, ecosystem, or earth level (Rolston III 2012).

In environmental ethics, values could be of different kinds and there has been an ongoing debate on a categorization of these values. One early attempt was David W. Ehrenfeld in 1976 and another was Holmes Rolston III in 1988. According to Ehrenfeld (1976) a non-resource could be ascribed a value and then turned into a resource so that it could be saved. Within this tradition, which according to Ehrenfeld originated with Aldo Leopold, some of these values are economic and others are non-economic. The values ascribed could for example be such as economic, recreational, aesthetic, historical, or cultural-symbolization value (Rolston III 1988). One can also argue through this categorization that economic value is one of many. A cumulative argument, that several other values combined could outweigh the economic value is possible. For Rolston III (1993), humans are the only species capable of being an ideal observer and as earlier quoted in Sect. 4.4 when discussing Biophilia, he claims humans can think freely beyond one's self-interests.

Another distinction in environmental ethics is between a dominion view and a stewardship view. The dominion view is tied to exploitation and overuse of resources, while the stewardship view is tied to an ethical careful usage of natural resources. There might also be a more anti-paternalistic position where humans are said to stay away from interfering with nature (cf. Regan 1983, 2013). This could be linked to the wilderness debate mostly conducted by American ethicists. When this book is written, one can hardly claim that true wilderness exists or that we can leave nature to itself. One example is that humans have directly or indirectly spread AMR globally also to places earlier considered wilderness (Sjölund et al. 2008) and human-induced climate change alters weather systems globally inflicting changes to ecosystems both on land and in water. There is a counterargument to the anti-paternalistic position that in fact more wild animals than expected depend on humans, and therefore humans need to take care of wildlife. Bjørn Ralf Kristensen (2023) uses an example from the Covid-19 lockdown where a reduction of visitors at a nature reserve (Stora Karlsö, at the east coast of Sweden) resulted in a decline in breeding success in the population of common guillemot due to more predation from white-tailed eagles than usual.

A further distinction is how to understand nature. One crucial question is whether culture is part of nature or culture is divided from nature. My position in the nature/culture discussion is that humans are a part of ecological systems regardless of how we live, either as people living close to nature or in cities. The fact that humans alter the environment is not unique. There is other so called keystone species that alter their environment. Examples of these are elephants clearing land and beavers making swamps. There are also several ethologists that talk about culture in animal species, especially in highly social mammals such as apes and elephants.

Although this is a diverse area, I will begin by discussing intergovernmental work in international bodies that protect nature by using a value system. This is particularly important because, as presented in Sect. 3.5.2, there have been proposals for a third kind of value, relational value, which is distinct from instrumental and intrinsic value. Then I analyze Aldo Leopold's land ethic which has been proposed as a fruitful ethic within the One Health approaches (see Table 3.1). Ecofeminism is part of environmental ethics, but I place this in Sect. 7.2.3. on pluralistic theories.

6.2.1 Intergovernmental Work on Relational Values: A Shift

Values in nature have been politically important as well as conceptually defined in the global strive to protect biodiversity and ecosystems. One important stakeholder in these matters is the Intergovernmental Platform on Biodiversity and Ecosystem Services (IPBES) and changes in their view of nature could give some additional insights into the discussion on where to go in One Health ethics. In Sect. 3.5.2, I mentioned that Foster et al. (2020) used the relational values proposed by IPBES to strengthen their view that ethics for Planetary Health need to extend rights to the more-than-human world. Therefore, I will analyze IPBES framework and the development of this kind of value.

In the framework of IPBES work for preserving biodiversity there has been a gradual shift from an earlier concept of ecosystem services to a concept of Nature's Contribution to People (NCP), and then another shift to their newest framework, the Nature Futures Framework (NFF). The concept of NCP is regarded as more inclusive than the original ecosystem services (Diaz et al. 2018; Ellis et al. 2019). The shift is intended to include more of social sciences and the humanities as well as Indigenous and local knowledge. By using the NCP-concept, it is believed that other values than instrumental economic ones could be added in the analysis. The concept is therefore more diverse in relation to values. These are described as "instrumental and non-instrumental relational values" (Ellis et al. 2019, 90).

However, in my view, the choice of term—nature's contribution to people—has an anthropocentric focus despite their argumentation that the new concept covers a diversity of values. I can't see that intrinsic values, which lack contribution to humans, could properly be included.

In their newest framework, NFF, the foundation is built on three kinds of values (Pereira et al. 2020; Kim et al. 2023). These are intrinsic values, relational values and instrumental values. Intrinsic values are labelled "nature for nature" and is defined as a "space allocated for nature" (Pereira et al. 2020, 1183) and "preserve nature for what it is" (Kim et al. 2023, 3). Relational values are labelled "nature as culture" and focus on "the relationships that nature and people co-create" (Kim et al. 2023, 3) and to be "one with nature" (Pereira et al. 2020, 1183). Instrumental values are labelled "nature for society" and resemble the old concept of ecosystem services (Pereira et al. 2020; Kim et al. 2023).

Relational values should here be seen as a distinct kind of value besides intrinsic and instrumental. This could alter the meta-ethical view on values, which often focuses on only two kinds of ethical value. Sandra Díaz et al. (2015) defines relational values as

> The values that are imbedded in desirable (sought after) relationships, including those among people and between people and nature; because such relationships are valued regardless of whether they imply tradeoffs to obtain *nature's benefits*, relational values depart from economic evaluation frameworks. (Díaz et al. 2015, 14, original emphasis)

Laura M. Pereira et al. further characterizes relational values (Nature as culture) and their meta-ethical foundation in the following quote:

> humans are perceived as an integral part of nature, and therefore what is valued is the reciprocal character of the people–nature relationship (Pereira et al. 2020, 1183)

Instrumental values are linked to benefits from nature that result in a good quality of life, while relational values focus on living in harmony with nature or the idea of Biophilia (see my discussion on the latter in Sect. 4.4). The difference between intrinsic values and relational is that intrinsic values are values not linked to human appreciation in any kind. This implies that both relational and instrumental values have this relation to humans, which is not the case for intrinsic values.

This development includes the possibility to work with biocentric and ecocentric frameworks in international conservation and preservation of ecosystems and species. This is because of the addition of intrinsic values (Nature for nature) within the framework. Also, non-western and non-colonial perspectives could be integrated because Indigenous peoples' views and livelihoods will be included within the relational values (Nature as culture) (see Kim et al. 2023).

6.2.2 Relational Values: Critique

Sanna Stålhammar and Henrik Thorén (2019) questions the position that relational value is a third kind of value, drawing on findings from three scientific disciplines (environmental ethics, ecosystem services valuation, and environmental psychology) that could use relational values. They think relational values might be a fruitful possibility in value theory, but they rather find that these values turn out to be descriptive when one studies how people value the environment rather than a new kind of theoretical value that could be used in ethics. However, I would like to contest this claim and try to find more firm foundation for relational values.

Although relational values seem to be exclusively developed within the field of protection of species and ecosystems, relational values exist more generally in philosophy, and they might be defined as

> *X* has *relational value* if and only if *X* has value in virtue of bearing some relation to something else. [… or …] *X* is valuable for *S* in virtue of being desired by *S* or being such that *S* would desire *X* were *S* "fully informed" and "rational." (Lemos 1999, 949, original emphasis)

Relational values such as they are defined in NFF relate to the first definition in the quote above. The relation itself is valuable (however not intrinsic), and it should be harmonious and reciprocal. Relational values also have importance for feminist ethics and indigenous ethics (Szende 2022).

6.2.3 *The Environment and Sustainability*

In environmental ethics, sustainability is a core concept. In Sect. 4.2.1 I introduced and discussed one version of the concept of sustainability, the Brundtland definition. That definition has been criticized for being too anthropocentric favoring economic value as more important than other values (Purvis et al. 2019). Therefore, another model often referred to as the three-pillar model of sustainability has been developed.[1] The model distinguishes sustainability as consisting of three areas, namely economic growth, environmental protection, and social progress. When the World Conservation Union (IUCN) discussed the three-pillar model they recognized that

> The environment is different [from the other two], since it is not created by society. (Adams 2006, 4)

The environment is supporting both economy and society and this analysis questions an earlier idea that tradeoffs could be between all three areas. By introducing what is called strong sustainability or critical natural capital one can restrict or disallow tradeoffs to protect ecosystems or species (Adams 2006). This makes the three-pillar model more biocentric or ecocentric than the Brundtland definition of sustainability.[2] The three-pillar model is in my view also compatible with Jean Lebel's (2003) view on the concept of health in EcoHealth as dependent on environment, economy, and community.

6.3 Land Ethic

Aldo Leopold's Land ethic is one framework within environmental ethics that has been proposed as a solution for One Health approaches (Goldberg and Patz 2015).[3] I will here analyze this framework which is mainly presented in his *A Sand County Almanac* published 1949 (Leopold 2013a) which is often also considered the start of environmental ethics. He argues that his ethics is the third phase in ethical development, where the first concerned relations between individuals, the second relations between individuals and human society, and the third between individuals and the surrounding land.

---

[1] The development of this model is somewhat unclear as has been evident elsewhere (Purvis et al., 2019).

[2] The choice of definition of sustainability also influences how one could interpret how to reach the Sustainable Development Goals.

[3] Indirectly, those persons arguing for Biophilia might also favor an ethic of Aldo Leopold because proponents for that perspective cites Leopold (see Sect. 4.4). However, Leopold's theory differs from the Biophilia perspective, and one could have one without the other.

In an earlier text, *The Conservation Ethic* (Leopold 2013f), he similarly develops the history of ethics in three phases, each of them with the focus on relations. Here, the third phase is yet to come and he states

> There is as yet no ethic dealing with man's relationship to land and to the non-human animals and plants which grow upon it. (Leopold 2013f, 326)

Leopold's ethics is relation-based and covers human relations to humans, animals, plants, and ecosystems.[4]

His Land Ethic developed in *A Sand County Almanac* was his final attempt to initiate the third phase.

One way to understand his land ethic is to start with his view of health. He defines health both on an individual human level as well as on an ecosystem level. In the section Wilderness in *A Sand County Almanac*, Aldo Leopold defines land health as an important concept (pp. 165–168) and he defines it in the section The Land Ethic:

> Health [of the land] is the capacity of the land for self-renewal. (Leopold 2013a, 185).[5]

The definition of health is similar for ecosystems (the land) and for humans in Leopold's writings, one just exchange "land" with "human".

The final section in *A Sand County Almanac*, The Land Ethic, is where he develops his final view on environmental ethics.[6] The land ethic has an individual conviction for being responsible for preserving this health and the medium how this is done is conservation. The latter term includes both the understanding of what constitutes for health as well as preservation of the same. What is Leopold's idea of individual ethics? Regarding hunting, Leopold, in his book on *Game Management*, writes that each hunter needs to formulate his own ethical code (Leopold 1933, 391). How does one decide whether an action is ethically right? Leopold answers in *A Sand County Almanac*:

> Examine each question in terms of what is ethically and esthetically right, as well as what is economically expedient. A thing is right when it tends to preserve the integrity, stability, and beauty of the biotic community. It is wrong when it tends otherwise. (Leopold 2013a, 188)

In an earlier publication, *The Ecological Conscience* from 1947, the same passage appears with some slight differences. The last sentence is not included and he continues the line that ends with "biotic community" with:

> … and the community includes the soil, waters, fauna and flora, as well as people. (Leopold 2013g, 531)

---

[4] This expansion of ethics is sometimes referred to as expanding circles of moral. Another proponent of this view has been Peter Singer in zoocentrism where animals are seen as the extension, see especially Singer (1983).

[5] Leopold (2013a, 165) defines also human health as the "capacity for internal self-renewal".

[6] He died April 21, 1948, and dated the Foreword March 4, 1948.

Leopold considers ethical and aesthetical values in relation to what it is possible to do within the economical frames. The ethical and aesthetical values are further defined in the concepts of integrity, stability, and beauty, which all three are applied to the level of the biotic community. He, therefore, has an ecosystem level of ethics. In the word community (also referred to as land), Leopold includes abiotic soil and water, and biotic plants and animals. That a possible economic value has a lower interest than ethical and aesthetic values are further explained as

> It is inconceivable to me that an ethical relation to land can exist without love, respect, and admiration for land, and a high regard for its value. By value, I of course mean something far broader than economic value; I mean value in the philosophical sense. (Leopold 2013a, 187)

In this quotation, a characterization of the relation between a human and the ecosystem is expressed. The relation should be based on love, respect and admiration.[7] However, the ethics proposed by Leopold in a *Sand County Almanac* are rather shallow.

By using his earlier works, one could provide deeper insight into his views on ethics although one must remember that he could have changed his views during time.

Another inherent ethical imperative, mentioned in some of his texts, is preservation between generations. As stated in his *The Virgin Southwest* (typescript from 1933):

> The major premise of civilization is that the attainments of one generation shall be available to the next. (Leopold 2013i, 317)

Leopold is modest, choosing only one next generation to consider (cf. Sect. 3.5.2 where three generations are suggested as sufficient). In his *The Conservation Ethic* from 1933 he develops that we are also restricted to what we got from earlier generations and the boundaries for us in the treatment of nature lie in the limits of the soil and the plant succession:

> We inherit the earth, but within the limits of the soil and the plant succession we also *rebuild* the earth (Leopold 2013f, 330, original emphasis)

When Fikret Berkes et al. (2012) analyzes Aldo Leopold's framework, they argue that Leopold proposes an ethical position of stewardship for the protection of ecosystems. To understand the entire picture of Leopold's legacy one must understand his holism. His rich understanding of ecology, human interactions with the environment, and his interest in creating policies and conservation of nature where humans still be a part of nature is needed to understand his ethics. Bryan G. Norton (1999) argues that Leopold's ethics was built on (for that time) robust empirical material

[7] Note here that the founders (Edward O. Wilson and Stephen R. Kellert) of the theory of Biophilia, which is described in Sect. 4.4, tries to use Leopold's idea of values and how he consider the relation of humans and nature to give strength to their theory. As was evident in Sect. 4.4, Biophilia is clearly anthropocentric in its valuation and Leopold (as argued here) is opening for intrinsic value in ecosystems.

(such as species knowledge, food web circuits, and different ways of dealing with nature) and then draw conclusions from this about possible ethical solutions. One could say that he was an empirical ethicist but tried to find a grand theory of how to explain all the things he saw. Norton also states that Leopold wasn't keen on theories standing alone from reality, he wanted to draw empirical conclusions from his theoretical thinking into real-world policies.

I will here continue to analyze each of the three concepts of integrity, stability, and beauty, which I (as well as Callicott (2022) and Radcliffe and Jessup (2022)) understand as the three most important concepts in his framework.

6.3.1 Integrity

For the concept integrity, he specifies this to physical integrity in his *Conservation Economics* from 1934 (Leopold 2013c, 343). But a more holistic view is presented in his *Some fundamentals of Conservation in the Southwest* (typescript from 1923):

> … we realize the indivisibility of earth—its soil, mountains, rivers, forests, climate, plants, and animals, and respect it collectively not only as a useful servant but as a living being, vastly less alive than ourselves in degree, but vastly greater than ourselves in time and space … (Leopold 2013e, 234)

Later, in his *A Survey of Conservation* (typescript from 1938) he defines the concept of conservation in a similar manner:

> Conservation is a state of harmony between man and land.
>
> By land is meant all of the things on, over, or in the earth.
>
> Harmony with land is like harmony with a friend; you cannot cherish his right hand and chop off his left … The land is one organism. Its parts, like our own parts, compete with each other and cooperate with each other. … You can regulate them—cautiously—but not abolish them. (Leopold 2013b, 416)

This also implies that one should not interfere with land areas that to our knowledge seem to function properly. The level of evaluation is on the whole system:

> If the land mechanism as a whole is good, then every part is good, whether we understand it or not. (Leopold 2013b, 416)

Wholeness and interconnectedness are therefore crucial for the understanding of integrity. Also, it is obvious that one could lack knowledge of parts of the ecosystem as long as one could draw a conclusion if the ecosystem works properly.

J. Baird Callicott, one of those that has studied and interpreted Leopold's framework in depth, argues that integrity is about biotic communities with a full setting of species composition (Callicott 2022).

6.3.2 Stability

The wholeness also connects to stability. A principle stated in *Lakes in Relation to Terrestrial Life Patterns* (A Symposium on Hydrobiology from 1941) on how to ethically evaluate technology was:

> A prudent technology should alter the natural order as little as possible. (Leopold 2013d, 453)

This could be understood in terms of stability. Natural systems should be as little interfered with as possible, and no overtaxing of ecosystems should be done. Leopold uses the example of soil sickness, which is a combination of loss of nutrients and erosion, to criticize American agriculture at this time as not being adapted to the local conditions. Leopold is critical of technology with the argument that technology further humans from nature. However, he sometimes sees nature as a machine and talks about humans as engineers of the natural world. So, he seems to relate to technology in different ways.

According to Callicott (2022), he argues that stability is linked to ecosystem stability, which has been seen as an equilibrium or as resilience depending on our knowledge of ecosystems. Radcliffe and Jessup (2022) argue that stability is the health of a diverse system of species.

6.3.3 Beauty

Beauty, the aesthetic value, is present through his writings. Beauty is distinguished from economic value, and I interpret that Leopold claims that species have aesthetic value regardless of their economic value. Still, it relates to economic value in an intricate relationship.

> The biota is beautiful collectively and in all its parts, but only a few of its parts are useful in the sense of yielding a profit to the private landowner. Healthy land is the only permanently profitable land, but if the biota must be whole to be healthy, and if most of its parts yield no saleable products, then we cannot justify ecological conservation on economic grounds alone. (Leopold 2013h, 516)

> The true problem of agriculture, and all other land-use, is to achieve both utility and beauty, and thus permanence. (Leopold 2013h, 517)

Beauty is connected to the wholeness of the ecosystem. In Leopold's *Game Management* he gives another definition of how to understand beauty. The strategy for preserving species should be:

> … to retain for the average citizen the opportunity to see, admire and enjoy, and the challenge to understand, the varied forms of birds and mammals indigenous to his state. It implies not only that these forms be kept in existence, *but that the greatest possible variety of them exist in each community*. (Leopold 1933, 403, original emphasis)

The greatest possible variety of species is therefore a guiding role to what wholeness implies in the beauty aspect. Daniel Simberloff (2012), in his analysis of the beauty idea in Leopold's writings, noted in the passage quoted above that foreign species, introduced or invasive, are excluded from the conservation aspect by being excluded from what is regarded as beauty. Callicott (2022) argues that beauty is explained as normal function on the ecosystem level and is therefore tied to land health, which is defined as capacity for self-renewal (see Sect. 6.3).

6.3.4 Towards a Land Ethic Decision Model

Does Leopold believe that his ideas of harmony with the land are reachable? In *A Survey of Conservation*, Leopold sees harmony with land as a somewhat utopic goal, but still a goal to strive for:

> We shall never achieve harmony with land, any more than we shall achieve justice or liberty for people. In these higher aspirations the important thing is not to achieve, but to strive. (Leopold 2013b, 423)

The main critique of the Land ethic is that it is vague. Although Leopold relies on the latest knowledge in ecology, his way of defining the ethical parts lacks the stringency for a modern ethical theory. Also, due to his death he didn't have the opportunity to develop his thoughts by discussing responses on the *Sand County Almanac*.

For me, I think there is an option to use the Land ethic and that is the reason that others have proposed this framework for One Health approaches. To be able to use it as a framework I shall try to adapt it to modern ethical claims. Leopold proposes a mixed framework. It is both virtue-based, focusing on love, respect and admiration as part of the virtues, and value-plural, acknowledging both ethical and aesthetic values of nature. As Radcliffe and Jessup (2022) notes, Leopold avoids providing principles on how to do ethics. He leaves to each human to decide for themselves on ethical matters. In order to be competent to do this, he clarifies the importance of educating ordinary people in (at that time) the new science of ecology as well as to reflect on one's conscience and moral development (Simpson and Cain 2000). In my view, he seems to contribute with a decision model. Let us go back to this quote:

> Examine each question in terms of what is ethically and esthetically right, as well as what is economically expedient. A thing is right when it tends to preserve the integrity, stability, and beauty of the biotic community. It is wrong when it tends otherwise. (Leopold 2013a, 188)

This could be transformed into a procedure where everyone with a proper knowledge level could frame the issue and ask oneself about the decisions. I will here explain a development I have done on Aldo Leopold's Land ethic, which I call The Land Ethic Decision Model. I will use that to analyze the cases because the model makes Leopold's theory more precise and easier to use. A modernization of an old theory that has been proposed although rarely applied due to its vagueness.

The Land Ethic Decision Model

The Land Ethic Decision Model consists of six questions based on Aldo Leopold's Land Ethic. The decision model starts with more particular aspects continuing to more holistic evaluations. Examine each question as holistic as possible for humans, animals, plants and ecosystems. This decision model meets the claims from Chap. 3 by Foster et al. (2020) for usefulness and Lederman (2018) for including humans, animals and the ecosystem.

1. **Does It Preserve the Beauty of the Biotic Community?**
   Here one can work with the principle that beauty for each ecosystem is the greatest possible biodiversity variety for each native lifeform. Focusing on native or indigenous species avoids problems that occur with non-native and invasive species. Notable in Aldo Leopold's thinking and in this decision-model is that particular individuals are less important than the greatest possible biodiversity variety at the species level.
2. **Does It Preserve the Stability of the Biotic Community?**
   Leopold's view on stability is the dynamics of ecosystem and a resilience view. Actions or decisions that risk overtaxing parts or the entire ecosystem should be avoided. On the earth level, this is also applicable to the Rockström and Gaffney (2021) view on planetary boundaries, that one should avoid tipping points. Here is also the place for intergenerational aspects to be included.
3. **Does It Preserve the Integrity of the Biotic Community?**
   Integrity is mainly a top-down principle. The most important guideline for integrity is if the entire ecosystem is functioning and sustainable in line with the Three-pillar model. All details might not be necessary if not available, unless there is a concern that this data deficiency might be crucial to understand the entire function of the ecosystem.
4. **What Is Economically Expedient?**
   When one considers whether decisions or actions is economically expedient, one should think about what is possible due to the economic model present related to forthcoming generations aspects as well as sustainability, which needs to be defined as the Three-pillar model presented in Sect. 6.2.3. Here also instrumental values such as ecosystem services or Nature's Contribution to People could be included.
5. **What Is Aesthetically Right to Do?**
   This question builds on the answer on question 1. Here, decisions or actions that destroy aesthetic aspects should be ruled out and avoided at all costs. This could be eradication of species or devastation of threatened ecosystems.
6. **What Is Ethically Right to Do?**
   This is the final question finding the overall decision possibilities. To decide what is ethically right to do, one works with values. Leopold's idea that the ethical relation must rely on philosophical value and no relation to land can exist without love, respect, and admiration indicates that philosophical values such as

intrinsic values and relational values mentioned in Sect. 6.2.1 should be considered. This inclusion of the modern way of analyzing values is strengthened by Leopold's urge to base his ethics on the newest available knowledge.

6.3.5 *Leopold: Bats Case*

Here follows the analysis of the Bats case by using the Land Ethic Decision Model.

Does it preserve the beauty of the biotic community?
Here one must consider the greatest possible biodiversity variety for each native lifeform. Eradication of bats or destroying their habitats, to minimize the risk of disease transmission, will definitely counteract this.

Does it preserve the stability of the biotic community?
Indigenous people's traditions must be preserved when they live in harmony with nature. Bat caves must be preserved. Cutting down forests risk overtaxing the ecosystems or altering the ecosystem completely (as what happens when rain forests are cut down). This might influence now living and forthcoming generations.

Does it preserve the integrity of the biotic community?
For the integrity of the biotic community, one could analyze the local society level. To work, there must be a preserved habitat to safeguard bat populations. Local livelihoods that have been depending on the relation with bats need to be assured to continue with this, and their sacred places should be preserved (all in line with IUCN SSC 2014).

What is economically expedient?
One of the calculations required to evaluate the economically expedient way to handle the issue is to weigh the benefits of deforesting with the potential costs of a dispersal of bats due to food shortages. This dispersal implies a greater risk of transferring diseases to humans as well as alter pollination habits which might impact other human resources.

What is aesthetically right to do?
Aesthetically wrong to do is to eradicate bats or destroy their habitats, such as cutting down the forests. That their habitat also is valued as sacred places influences how one values the aesthetic aspect.

What is ethically right to do?
This is an overall estimation of what has been seen in the other questions. Preserving local livelihoods which for an extended period of time co-existed with bats seem to be ethically right to do consider all aspects of the case.

6.3.6 Leopold: Animal Assisted Therapy Case

Here follows the analysis of the Animal Assisted Therapy case by using the Land Ethic Decision Model.

Does it preserve the beauty of the biotic community?
If one adjusts the therapy so that damage on vegetation is as much avoided as possible together with taking care of pollution and manure, there will not be a great risk of altering biodiversity, even though one is outdoors. Riding outdoors in nature might facilitate human appreciation of the beauty of the ecosystems.

Does it preserve the stability of the biotic community?
One issue here to compensate for is the risk of overgrazing by horses with the result of altering the biotic community's carrying capacity. One must have enough land for the number of horses involved, to avoid that the carrying capacity of the ecosystem is overtaxed and altered to lower levels for our or future generations.

Does it preserve the integrity of the biotic community?
As long as this therapy is environmentally friendly, I cannot see that there might be issues that risk the integrity of the biotic community.

What is economically expedient?
An economic analysis on costs and benefits could quite easily be done on all those affected in and by the therapy. One problem generally with equine assisted therapy is that it tends to have high costs in the short run (during therapy). If one considers the long run, the rest of a lifetime, a change in the way of life might be an economic benefit. Changes in the environment due to this practice such as excess of manure, euthrophication of water, overtaxing of vegetation (see analysis on stability) should also be considered at the ecosystem level.

What is aesthetically right to do?
Taken together this seem to preserve the beauty of the environment.

What is ethically right to do?
Most of what the Land ethic focuses on is somewhat outside the issues in this case. Therefore, this is a case which quite easily could be analyzed by the Land ethic and compensating for the ecological risks, the AAT case can be ethically sound.

6.3.7 Leopold: Oceans Case

Here follows the analysis of the Oceans case by using the Land Ethic Decision Model.

Does it preserve the beauty of the biotic community?
Here one must consider the greatest possible biodiversity variety for each native lifeform. We know that the effects of climate change could destroy coral reefs with the risk of extinction of many species. Rising sea levels alter the coastline rapidly

with a change of habitats. If, for example, mangrove forest cannot follow the changing coastal lines quickly enough, many species may be extinct. Therefore, every effort to minimize climate change needs to be performed to not lose Leopold's idea of beauty.

For overfishing, the situation is somewhat different. To preserve the greatest possible biodiversity, all nations need to follow the decided fishing quotas. Companies involved in fishing need to consider methods and processes that results in less waste at the fishing grounds (i.e. other species killed that is not aimed for in the catch) as well as in the manufacturing of products (so that less fish needs to be caught).

Does it preserve the stability of the biotic community?
The main threat to oceans, their species (especially in the coastal regions) and humans are twofold when temperatures alter; heatwaves could warm surface water risking extinction, and rising sea levels turn terrestrial land into coastal zones altering the dynamics and resilience for the ecosystems. This will also influence future generations.

Does it preserve the integrity of the biotic community?
Concerning the top-down view on the biotic community there need to be a core focus on overarching aspects. One of these might be preventing pollution. Pollution brings chemical compounds and microplastics to water which influences many species' lives. Chemicals might be transferred within the ecosystem webs until they have been aggregated as much as being lethal. Therefore, pollution should be avoided as much as possible.

Decisions that tend to ruin local livelihoods living in harmony with nature are also a top-down aspect. Then one lose biodiversity as well as culture at the same time.

What is economically expedient?
Here one must do a total economic analysis also including values that are not easy to turn into economic values, an issue already mentioned by Ehrenfeld (1976; Sect. 6.2). One also must do trade-offs between different solutions to see where the economic costs and benefits would be. This must be evaluated for the long term, most probably up to three generations in advance.

What is aesthetically right to do?
Aesthetically one should strive to preserve as much biodiversity as possible, especially given the ongoing eradication of entire populations and whole species in the oceans due to climate change, pollution, and over-fishing.

What is ethically right to do?
This is an overall estimation of what has been seen in the other questions. For overfishing, this might imply that one might have a more global approach to distribution of food resources (both aquatic and terrestrial) to avoid overfishing and biodiversity depletion. Due to the high risk of losing stability of the biotic community and the beauty of the biotic community when climate warming is continuing, the ethically right thing to do must be to lower the emissions.

6.3.8 *Conclusion Land Ethic*

Leopold's land ethic works well in the Bats case and the Oceans case but less well in the Animal Assisted Therapy case. Even if the theory is a little thin in Leopold's own writings, my proposed decision model might work fairly well. However there is some overlap in the aspects of Leopold's thoughts which might need further deliberation to be easier to use.

6.4 Summary

In this chapter, ethical frameworks in environmental ethics have been analysed. The work by IPBES has entered a new kind of values into the debate, relational values. They need to be included in the ethical discussion, and I have tried to analyse what they are and how to use them. Aldo Leopold's framework is fruitful for One Health approaches, and I have developed part of it to a Land Ethic Decision Model to better fit the ethical imperative from the One Health approaches. Next chapter will deal with pluralistic ethics.

References

Adams, W.M. 2006. The future of sustainability: Re-thinking environment and development in the twenty-first century. Report of the IUCN Renowned Thinkers Meeting, 29-31 January 2006. https://portals.iucn.org/library/sites/library/files/documents/Rep-2006-002.pdf

Berkes, Fikret, Nancy C. Doubleday, and Graeme S. Cumming. 2012. Aldo Leopold's land health from a resilience point of view: Self-renewal capacity of social-ecological systems. *EcoHealth* 9:278–287.

Callicott, J. Baird. 2022. Three project ideas to fill research gaps in Aldo-Leopold scholarship. *Socio-Ecological Practice Research* 4:11–16.

Díaz, Sandra, Sebsebe Demissew, Julia Carabias, Carlos Joly, Mark Lonsdale, Neville Ash, Anne Larigauderie, Jay Ram Adhikari, Salvatore Arico, András Báldi, Ann Bartuska, Ivar Andreas Baste, Adem Bilgin, Eduardo Brondizio, Kai M. A. Chan, Viviana Elsa Figueroa, Anantha Duraiappah, Markus Fischer, Rosemary Hill, Thomas Koetz, Paul Leadley, Philip Lyver, Georgina M. Mace, Berta Martin-Lopez, Michiko Okumura, Diego Pacheco, Unai Pascual, Edgar Selvin Pérez, Belinda Reyers, Eva Roth, Osamu Saito, Robert John Scholes, Nalini Sharma, Heather Tallis, Randolph Thaman, Robert Watson, Tetsukazu Yahara, Zakri Abdul Hamid, Callistus Akosim, Yousef Al-Hafedh, Rashad Allahverdiyev, Edward Amankwah, Stanley T. Asah, Zemede Asfaw, Gabor Bartus, L. Anathea Brooks, Jorge Caillaux, Gemedo Dalle, Dedy Darnaedi, Amanda Driver, Gunay Erpul, Pablo Escobar-Eyzaguirre, Pierre Failler, Ali Moustafa Mokhtar Fouda, Bojie Fu, Haripriya Gundimeda, Shizuka Hashimoto, Floyd Homer, Sandra Lavorel, Gabriela Lichtenstein, William Armand Mala, Wadzanayi Mandivenyi, Piotr Matczak, Carmel Mbizvo, Mehrasa Mehrdadi, Jean Paul Metzger, Jean Bruno Mikissa, Henrik Moller, Harold A. Mooney, Peter Mumby, Harini Nagendra, Carsten Nesshover, Alfred Apau Oteng-Yeboah, György Pataki, Marie Roué, Jennifer Rubis, Maria Schultz, Peggy Smith, Rashid Sumaila, Kazuhiko Takeuchi, Spencer Thomas, Madhu Verma, Youn Yeo-Chang, and

Diana Zlatanova. 2015. The IPBES conceptual framework: Connecting nature and people. *Current Opinion in Environmental Sustainability* 14:1–16.
Díaz, Sandra, Unai Pascual, Marie Stenseke, Berta Martín-López, Robert T. Watson, Zsolt Molnár, Rosemary Hill, Kai M. A. Chan, Ivar A. Baste, Kate A. Brauman, Stephen Polasky, Andrew Church, Mark Lonsdale, Anne Larigauderie, Paul W. Leadley, Alexander P. E. van Oudenhoven, Felice van der Plaat, Matthias Schröter, Sandra Lavorel, Yildiz Aumeeruddy-Thomas, Elena Bukvareva, Kirsten Davies, Sebsebe Demissew, Gunay Erpul, Pierre Failler, Carlos A. Guerra, Chad L. Hewitt, Hans Keune, Sarah Lindley, and Yoshihisa Shirayama. 2018. Assessing nature's contributions to people. *Science* 359 (6373): 270–272.
Ehrenfeld, David W. 1976. The conservation of non-resources. *American Scientist* 64:648–656.
Ellis, Erle C., Unai Pascual, and Ole Mertz. 2019. Ecosystem services and nature's contribution to people: Negotiating diverse values and trade-offs in land systems. *Current Opinion in Environmental Sustainability* 38:86–94.
Foster, Alexander, Jennifer Cole, Ivica Petrikova, Andrew Farlow, and Howard Frumkin. 2020. Planetary health ethics. In *Planetary health: Protecting nature to protect ourselves*, ed. Samuel Myers and Howard Frumkin, 453–473. Washington and Covelo: Island Press.
Goldberg, Tony L., and Jonathan A. Patz. 2015. The need for a global health ethic. *Lancet* 386:e37–e39. https://doi.org/10.1016/S0140-6736(15)60757-7.
IUCN SSC. 2014. IUCN SSC Guidelines for minimizing the negative impact to bats and other cave organisms from guano harvesting. Ver. 1.0. IUCN, Gland.
Kim, Hye Jin, Garry D. Peterson, William W. L. Cheung, Simon Ferrier, Rob Alkemade, Almut Arneth, Jan J. Kuiper, Sana Okayasu, Laura Pereira, Lilibeth A. Acosta, Rebecca Chaplin-Kramer, Eefje den Belder, Tyler D. Eddy, Justin A. Johnson, Sylvia Karlsson-Vinkhuyzen, Marcel T. J. Kok, Paul Leadley, David Leclère, Carolyn J. Lundquist, Carlo Rondinini, Robert J. Scholes, Machteld A. Schoolenberg, Yunne-Jai Shin, Elke Stehfest, Fabrice Stephenson, Piero Visconti, Detlef van Vuuren, Colette C. C. Wabnitz, Juan José Alava, Ivon Cuadros-Casanova, Kathryn K. Davies, Maria A. Gasalla, Ghassen Halouani, Mike Harfoot, Shizuka Hashimoto, Thomas Hickler, Tim Hirsch, Grigory Kolomytsev, Brian W. Miller, Haruka Ohashi, Maria Gabriela Palomo, Alexander Popp, Roy Paco Remme, Osamu Saito, U. Rashid Sumalia, Simon Willcock, and Henrique M. Pereira. 2023. Towards a better future for biodiversity and people: Modelling nature futures. *Global Environmental Change* 82:102681.
Kristensen, Bjørn Ralf. 2023. Anthrodependency, zoonoses, and relational spillover. In *More-than-one-health: Humans, animals, and the environment post-Covid*, ed. Irus Braverman, 193–208. London and New York: Routledge.
Lebel, Jean. 2003. Health: An ecosystem approach focus. International Development Research Centre: Ottawa, ON. Available from: https://www.idrc.ca/en/book/infocus-health-ecosystem-approach
Lederman, Zohar. 2018. *Stamping out culling: The ethics of culling from a one health perspective*. Ph. D. Thesis. Centre for Biomedical Ethics, National University of Singapore.
Lemos, Noah M. 1999. Value. In *The Cambridge dictionary of philosophy*, ed. Robert Audi, 2nd ed., 948–949. Cambridge and New York: Cambridge University Press.
Leopold, Aldo. 1933. *Game management*. New York: Charles Scribner's Sons.
Leopold, Aldo. 2013a. A sand county almanac. In *Aldo Leopold: A Sand County Almanac & other writings on ecology and conservation*, ed. Curt Meine, 1–189. New York: The Library of America.
Leopold, Aldo. 2013b. A survey of conservation. (Typescript). In *Aldo Leopold: A Sand County Almanac & other writings on ecology and conservation*, ed. Curt Meine, 416–424. New York: The Library of America.
Leopold, Aldo. 2013c. Conservation economics. In *Aldo Leopold: A Sand County Almanac & other writings on ecology and conservation*, ed. Curt Meine, 340–351. New York: The Library of America.
Leopold, Aldo. 2013d. Lakes in relation to terrestrial life patterns. In *Aldo Leopold: A Sand County Almanac & other writings on ecology and conservation*, ed. Curt Meine, 447–453. New York: The Library of America.

Leopold, Aldo. 2013e. Some fundamentals of conservation in the southwest. In *Aldo Leopold: A Sand County Almanac & other writings on ecology and conservation*, ed. Curt Meine, 224–236. New York: The Library of America.

Leopold, Aldo. 2013f. The conservation ethic. In *Aldo Leopold: A Sand County Almanac & other writings on ecology and conservation*, ed. Curt Meine, 325–339. New York: The Library of America.

Leopold, Aldo. 2013g. The ecological conscience. In *Aldo Leopold: A Sand County Almanac & other writings on ecology and conservation*, ed. Curt Meine, 524–532. New York: The Library of America.

Leopold, Aldo. 2013h. The land-health concept and conservation. In *Aldo Leopold: A Sand County Almanac & other writings on ecology and conservation*, ed. Curt Meine, 512–518. New York: The Library of America.

Leopold, Aldo. 2013i. The virgin southwest. In *Aldo Leopold: A Sand County Almanac & other writings on ecology and conservation*, ed. Curt Meine, 317–324. New York: The Library of America.

Norton, Bryan G. 1999. Leopold as practical moralist and pragmatic policy analyst. In *The essential Aldo Leopold: Quotations and commentaries*, ed. Curt Meine and Richard L. Knight, 201–219. Madison, WI: The University of Wisconsin Press.

Pereira, Laura M., Kathryn K. Davies, Eefje den Belder, Simon Ferrier, Sylvia Karlsson-Vinkhuyzen, Hyejin Kim, Jan J. Kuiper, Sana Okayasu, Maria G. Palomo, Henrique M. Pereira, Garry Peterson, Jyothis Sathyapalan, Machteld Schoolenberg, Rob Alkemade, Sonia Carvalho Ribeiro, Alison Greenaway, Jennifer Hauck, Nicholas King, Tanya Lazarova, Federica Ravera, Nakul Chettri, William W. L. Cheung, Rob J. J. Hendriks, Grigoriy Kolomytsev, Paul Leadley, Jean-Paul Metzger, Karachepone N. Ninan, Ramon Pichs, Alexander Popp, Carlo Rondinini, Isabel Rosa, Detlef van Vuuren, and Carolyn J. Lundquist. 2020. Developing multiscale and integrative nature-people scenarios using the natures futures framework. *People and Nature* 2:1172–1195.

Purvis, Ben, Yong Mao, and Darren Robinson. 2019. Three pillars of sustainability: In search of conceptual origins. *Sustainability Science* 14:681–695.

Radcliffe, Robin W., and David A. Jessup. 2022. Wildlife health and the North American model of wildlife conservation. *Journal of Zoo and Wildlife Medicine* 53 (3): 493–503.

Regan, Tom. 1983. *The case for animal rights*. Berkeley: University of California Press.

Regan, Tom. 2013. Animal rights and environmental ethics. In *The structural links between ecology, evolution and ethics: The virtuous epistemic circle. Boston studies in the philosophy of science*, ed. D. Bergandi, vol. 296, 117–126. Dordrecht: Springer.

Rockström, Johan, and Owen Gaffney. 2021. *Breaking boundaries: The science of our planet*. New York: DK Publishing, Penguin Random House.

Rolston, Holmes, III. 1988. *Environmental ethics: Duties to and values in the natural world*. Philadelphia: Temple University Press.

Rolston, Holmes, III. 1993. Biophilia, selfish genes, shared values. In *The Biophilia hypothesis*, ed. Stephen R. Kellert and Edward O. Wilson, 381–414. Washington, DC: Island Press.

Rolston, Holmes, III. 2012. *A new environmental ethics: The next millennium for life on earth*. New York and London: Routledge.

Ruggiero, Michael A., Dennis P. Gordon, Thomas M. Orrell, Nicolas Bailly, Thierry Bourgoin, Richard C. Brusca, Thomas Cavalier-Smith, Michael D. Guiry, and Paul M. Kirk. 2015. A higher level classification of all living organisms. *PLoS One* 10 (4): e0119248. https://doi.org/10.1371/journal.pone.0119248.

Simberloff, Daniel. 2012. Integrity, stability, and beauty: Aldo Leopold's evolving view of nonnative species. *Environmental History* 17 (3): 487–511.

Simpson, Steven V., and Kelly D. Cain. 2000. Recreation's role in the environmental ethics dialogue: The case of Aldo Leopold and the morality of hunting. *Leisure/Loisir* 25(3–4): 181–197. [Reprinted, originally published in Journal of Applied Recreation Research 1995, 20(3): 185–201.]

Singer, Peter. 1983. The expanding circle: Ethics and sociobiology. Paperback edition. Oxford and Melbourne: Oxford University Press.

Sjölund, Maria, Jonas Bonnedahl, Jorge Hernandez, Stina Bengtsson, Gunilla Cederbrant, Jarone Pinhassi, Gunnar Kahlmeter, and Björn Olsen. 2008. Dissemination of multidrug-resistant bacteria into the Arctic. *Emerging Infectious Diseases* 14 (1): 70–72.

Stålhammar, Sanna, and Henrik Thorén. 2019. Three perspectives on relational values of nature. *Sustainability Science* 14:1201–1212.

Szende, Jennifer. 2022. Relational value, land, and climate justice. *Journal of Global Ethics* 18 (1): 118–133.

Wägner, Elin. 2007[1941]. Väckarklocka. Stockholm: Albert Bonniers Förlag [In Swedish].

Chapter 7
A Need for a Pluralistic Ethics

Rather we aim to be allies, supporting the clans in their travel along sustainable development pathways of their own design. (Middleton et al. 2020, 1659)

Abstract In this chapter, the final of the four analyzing chapters, pluralistic ethical frameworks are studied. Although pluralistic ethics have been suggested within One Health approaches, no framework has been proposed. After a brief introduction, I will give a short overview of pluralistic ethics by pinpointing some interesting frameworks that might be helpful. In the overview, I therefore add Onora O'Neill's idea of bridging virtue-based ethics with deontological ethics as a fruitful example. I also analyze values in Indigenous ethical codes that could be helpful on what values to include when one includes knowledge systems different from western science. Main emphasis in this chapter will be on the dialogue ethics proposed by Seyla Benhabib, who works with the most fruitful framework hitherto in bridging cultures and religions into ethical deliberation. Although her framework is good; I recognize it to be further developed in the pre-framing phase of the dialogue. I also incorporate the Ubuntu theory of an individual's relation to others, with its emphasis on including an individual's enemies. Finally, I develop my own model for an ethical dialogue. Which is a two-stage dialogue, where the first stage is how to frame the ethical dialogue. The first stage is needed to gather as many perspectives as needed to solve ethical cases. In this stage, one must discuss inherent assumptions due to cultural and/or religious differences. The first dialogue sets the agenda and chooses participants for the second dialogue, which then will be an open-ended dialogue which similarities to Benhabib's suggestions within dialogue ethics. The result of the second dialogue is ethical solutions to the case and if the procedure has been fair in both dialogues, the procedure is ethically good.

H. Lerner, *Ethics for One Health Approaches: A Roadmap for Future Directions*, Advancing Global Bioethics 22,
https://doi.org/10.1007/978-3-032-05239-1_7

Keywords Ecofeminism · Onora O'Neill · Seyla Benhabib · Indigenous Peoples · Ubuntu · Dialogue ethics · Discourse ethics · two-stage dialogue · First Nations

7.1 Introduction

This final chapter of analyzing ethical frameworks proposed within the One Health approaches will concern pluralistic ethical frameworks. The procedure is similar. After the overview of the field, each proposed framework will be presented in its original form and then developed or criticized depending on content or suitability to One Health approaches, and thereafter applied to the three cases presented in Sect. 1.4. This application will indicate strengths and weaknesses of the framework analyzed.

7.2 Overview of Pluralistic Ethics

Previous chapters have established that promising frameworks of ethics that could be used for One Health approaches unfortunately might be too narrow to include pluralistic values and perspectives. Western ethics has had a struggle finding bridges between the major ethical theories (utilitarianism, deontological ethics, virtue ethics, contractarian ethics, and feminist ethics) although there are attempts made. There are also attempts to bridge the big strains of ethics based on moral status, namely anthropocentrism, zoocentrism, biocentrism, and ecocentrism. Ethics originating from non-western and indigenous scholars is also important to bridge to be able to reach a promising ethic for One Health approaches (see Chap. 2). Seyla Benhabib frames the problem that has been hampering ethics in the following quote:

> At the center of this debate is the question: if what had hitherto been considered the major works of the western tradition are, almost uniformly, the product of a specific group of individuals, namely propertied, white, European and North American males, how universal and representative is their message, how inclusive is their scope, and how unbiased their vision? (Benhabib 1992, 242)[1]

To overcome this huge exclusion of views and perspectives possible, a pluralistic account and/or a more cultural- or context-open ethical framework might be needed. Following the epigraph by Jo Middleton et al. (2020), we should try to find bridges and connections to develop such a new framework of ethics already started. Aware that a single chapter in a book will not give proper justice to all the interesting

[1] Benhabib's point is crucial in each writer's self-reflection on the context where the writer works. I fall within this quotation, being a white, western male. I hope that this book will try to be inclusive outside my own context, but that will be up to the reader to decide. I fully agree with Benhabib that in order to find an ethical pathway for One Health one need to change the discussion to be as inclusive as possible.

attempts being made, I will here address some of the most promising. I will start with important aspects of pluralism from Onora O'Neill. Then, I turn to ethical discussions on research ethics where Indigenous peoples are included and continue with critique from ecofeminism. This is mainly to meet the demand that One Health ethics must be gender-sensitive, intersectional, interreligious, non-western, and non-colonial. The Ebola outbreak in 2014-2015 in West Africa is one important example. Several scholars analyzing this outbreak stress the importance of including local religious leaders (Manguvo and Mafuvadze 2015; Marshall 2017). If so, one could more easily reach out to the local communities.

> Reviews of the Ebola experience should highlight the important roles of religious actors, above all because of their dense community networks and frequent high levels of trust. (Marshall 2017, 632)

The main framework to analyze in this chapter is the dialogue ethics of Seyla Benhabib, which will be discussed at length. I problematize her theory with the African concept of Ubuntu analyzed by Michael Onyebuchi Eze in order to develop her theory. Finally, I present my own suggestion for a two-stage dialogue that will meet the claims of inclusiveness early in the process and take care of the risk of losing perspectives when opposing views avoid meeting each other at the table. I think that my suggestion opens a path to reaching agreement on how to form a dialogue (stage 1) as well as perform the dialogue (stage 2). For the second stage I rely heavily on Benhabib.

7.2.1 A Constructivist Account of Ethics: The Blueprint

Philosopher Onora O'Neill (1996) analyzes what a constructive account of ethics might be able to do. Instead of arguing for an ethics that prescribes details in judgments as a result of applying an ethical theory, her view is that a constructive account of pluralistic ethics is open and works as a blueprint.

> Like other blueprints, they will specify structural requirements, but leave other matters undetermined. (O'Neill 1996, 209)

And continues later at the end of her book:

> What it offers is a reasoned way of thinking about the planning and construction of earthly cities and of lives that are or could be led and shared in them under various conditions without fantasizing that imaginary foundations are available or that permanent solutions are possible. (O'Neill 1996, 211)

This down-to-earth guiding version has less precision than many ethical theories because the "blueprint" only gives a guiding example of how to do. Compared to other theories, more is left to the ethical deliberation in the present case to settle the issue. This widens the ways an ethical case can be solved. O'Neill compare herself with John Rawls' theory (as a foundation for wide reflective equilibrium) and finds that when Rawls focus on "plurality of beliefs", she focuses on "plurality of agents"

(O'Neill 2013). For me this is more useful in One Health approaches where one cannot suppose that all agents share similar values or ethical systems.

Her own constructive account of ethics blends justice, as a principle-based modified version of Kant's categorical imperative,[2] with virtue-based ethics. Justice is focused on an obligation to reject direct and indirect injury, especially the systematic and gratuitous one. Virtues are focused on an obligation to reject indifference and neglect. In her work (O'Neill 1996) ethics are multilayered. Where you start, the top-layer is justice and then you continue with the second layer, virtues. Finally, in the third layer, you are allowed to solve cases differently due to other relevant matters. Here, there is no guidance, hence, the open idea of her ethics. This could be further developed in her reference to a blueprint. Consider a villa area and its blueprint. There could be three layers, the entire villa area, the villa plot and the villa itself, and inside the villa. For all citizens living in this area, the blueprint decides where the different plots are, so that water, power and roads could be provided. Each house and plot need to follow some restrictions in order to not annoy other neighbors. This could be the allowed height of the house, how the roof must be constructed, rules for where kitchen must be placed etcetera. Finally, how decorating certain rooms is up to each family.

The point with O'Neill's pluralistic ethic is to provide guidelines, not offering complete instruction (O'Neill 1996, 212).

Both in the rejection of indirect injury as well as the rejection of indirect indifference and neglect, natural environments are included (O'Neill 1996). She avoids further development but exemplifies that preservation and conservation of natural environments are covered by her framework.

7.2.2 Values in Indigenous Ethical Codes

In the analysis of One Health approaches in Chap. 2 it became evident that all three major One Health approaches (One Health, EcoHealth, and Planetary health) were able to acknowledge Indigenous peoples' knowledge as important as Western science. In the Convention of Biological Diversity, Article 8j, (adopted 1992) Indigenous knowledge and local communities' knowledge has a specific protection if their knowledge concerns biodiversity:

> Subject to its national legislation, respect, preserve and maintain knowledge, innovations and practices of indigenous and local communities embodying traditional lifestyles relevant for the conservation and sustainable use of biological diversity and promote their wider application with the approval and involvement of the holders of such knowledge, innovations and practices and encourage the equitable sharing of the benefits arising from the utilization of such knowledge, innovations and practices. (Secretariat of the Convention on Biological Diversity 2011, Article 8j, 8)

---

[2] She avoids the categorical imperative and have a more open principle of avoiding direct and indirect injury to others.

The article gives holders of the knowledge a specific role. They are granted participation and the right to prove or disapprove of the use of their knowledge depending on if they get an equitable sharing of benefits or not. Therefore, differences in procedures and values between western scientific ethics and Indigenous People's ethics has been studied in policy making with the aim to find policies as well as procedures to include both sides reciprocal and fair. Several scholars—Herman (2014), Tunón et al. (2016), and Brant et al. (2023)—have analyzed ethical codes, guidelines and frameworks especially in the field of research ethics, where the aim has been inclusive to both groups in order to find a common ethical foundation. Included in all three studies, well above 40 documents were considered with a minor overlap. In all three studies, three ethical concepts are present. These are

- Respect
- Responsibility
- Reciprocity

Respect is favoring a just and equitable process with recognition and alignment with Indigenous rights, worldviews, and values.

Responsibility concerns the researcher and how confidentiality and consent are handled, but also protection of knowledge so that Indigenous Peoples avoid suffering from for example biopiracy.

Reciprocity builds on partnership with mutual benefits and relational accountability. In the Akwé: Kon guidelines (Secretariat of the Convention on Biological Diversity 2004) reciprocity is understood as involving all aspects of preparatory work, planning, development, and implementation of a project.

7.2.3 Ecofeministic Critique

Another suitable perspective of inclusive thinking originates from ecofeminism. I will here briefly introduce ecofeminism and try to briefly present the main lines that might be important for One Health ethics as a part of the roadmap.

Karen J. Warren (2015) claims that ecofeminism, as being part of feminist ethics, has a focus on both criticizing male-dominated ethics and favor non-male-dominated ethics. The focus on values in ecofeminism is on care, empathy, and friendship—values less discussed in Western ethics—and the framework has focus on connections between women and nature. Although, extremely diverse as a field, she summarizes ecofeminist philosophical ethics in four themes. First, ecofeminism questions the dualism between nature and culture. In ecofeminism, one person could both have an individual self, separated from nature, as well as an ecological self that is connected to nature. Therefore, it follows secondly that those selves are relational and therefore relationships in itself have values, and the nature of the relationship matters morally (compare this to the discussion of IPBES values in Sect. 6.2.1). Third, ecofeminism is contextual and due to historical and cultural differences. It needs a dialogical form that is based on procedure rather than principles, rules or rights. There is a strong claim for a shift from monism to pluralism in ethical decision-making and expression. Fourth, ecofeminism claims that this diversity of connections between woman and nature and its focus on gender bias is better than other theories and that there is useless to find an objective moral point of view.

But in my analysis, I don't take ecofeminism as a main road because I believe Seyla Benhabib's dialogue ethics manage to incorporate the feminist, anti-colonial and posthumanistic claim in Chap. 3. I will therefore clarify and claim that Benhabib covers these issues within her theory.

7.3 Pluralistic Methods: Dialogue Ethics

In political philosophy, discourse ethics or dialogue[3] ethics has been an important strand focusing on procedural rather than principled ethics. Among the most prominent has been Karl-Otto Apel, Jürgen Habermas, and Seyla Benhabib. Where Apel and Habermas has focused more on what the outcome of the dialogue could be, Benhabib has focused on the procedure itself. I will here argue that Benhabib's version of dialogue ethics is the most suitable framework to meet the claims from Chap. 3 that One Health ethics need a framework that encompass feminist, anti-colonial (including Indigenous ethics), posthumanistic, and pluralistic ethics.

Benhabib has her philosophical roots in feminism and discourse ethics, proposing a dialogue ethics that might be the widest pluralistic version of these at the moment. Her framework focuses on gender and cultural discrimination as well as heterogeneity in society, which meets some of the barriers that the Network for Ecohealth and One Health (NEOH) found (Laing et al. 2023; see Sect. 3.7). It might also meet claims from Indigenous peoples. Therefore, this approach will be interesting to analyze. I mainly develop her thoughts from her book *Situating the Self* (Benhabib 1992) and her essay *Toward a deliberative model of democratic legitimacy* (Benhabib 1996b).

> All argumentation entails respect for one's conversation partners; such respect belongs to the idea of fair argumentation; to be a competent partner in such a conversation then entails recognizing the principle of equal respect. (Benhabib 1992, 31)

A dialogue needs to have at least two partners in order to be a dialogue. A fair argumentation needs to have an equal standing by the people involved. Compared to the suggestion by Foster et al. (2020) in Planetary Health (see Sect. 3.5.2), this is a more proper way of considering power relations. In Foster et al. dialogue seems to be one-sided when the Planetary Health educated people give the information and arrange for others included how to respond to their information. There is a greater risk for

[3]This strand of philosophy has had different names. Sometimes it is called communicative ethics, discourse ethics, or dialectical ethics. I will in this chapter use dialogue ethics as a term because it is the best fit for Seyla Benhabib's framework.

lack of equal respect due to this power relation in one side choosing the terms for the discussion. This is not the case in Benhabib's approach to dialogue.

Benhabib questions all ethical attempts based on ethical principles (such as utilitarianism and deontology) including all forms of principlism.

> The core intuition behind modern universalizability procedures is not that everybody could or would agree to the same set of principles, but that these principles have been adopted as a result of a procedure, whether of moral reasoning or of public debate, which we are ready to deem "reasonable and fair". (Benhabib 1992, 37)

Although she refers to intuition, that intuition is based on what she considers to be the case when one needs to find an ethic for multicultural societies. As stated elsewhere, she wrote that diversity in society cannot be solved by eliminating otherness and difference creating a homogenization, which has been the case in many western liberal societies (Benhabib 1996a). She highlights the risk of missing crucial information in decision-making, questioning the notion that one person can find the solution to a problem.

> New information is imparted because 1) no single individual can anticipate and foresee all the variety of perspectives through which matters of ethics and politics would be perceived by different individuals; and 2) no single individual can possess all the information deemed relevant to a certain decision affecting all. (Benhabib 1996b, 71)[4]

She is aiming for an open-ended process of ethical deliberation.

> … the processual generation of reasonable agreement about moral principles via an open-ended moral conversation. (Benhabib 1992, 37)

To be an open-ended conversation its result must be reached by a reasonable and fair deliberation. She holds that this process is a continuous process:

> … the dialogic model of ethics defended … envisages a continuous process of conversation in which understanding and misunderstanding, agreement as well as disagreement are intertwined and always at work. (Benhabib 1992, 197–198)

She further develops this continuous open-ended process in the following quote:

> … 1) participation in such deliberation is governed by the norms of equality and symmetry; all have the same chances to initiate speech acts, to question, to interrogate, and to open debate; 2) all have the right to question the assigned topics of conversation; and 3) all have the right to initiate reflexive arguments about the very rules of the discourse procedure and the way in which they are applied and carried out. There are no prima facie rules limiting the agenda of the conversation, or the identity of the participants, as long as each excluded person or group can justifiably show that they are relevantly affected by the proposed norm under question. (Benhabib 1996b, 70)

One core competence that a person should have is the ability of "reversibility of perspectives". She defines it as

> … we practice the reversibility of perspectives implicit in adult human relationships. The development of this capacity for reversing perspectives and the development of the capacity

[4] This is of course also a relevant critique of this book. How objective and inclusive I intend to be, this book is still very much a result of my own conclusions.

> to assume the moral point of view are intimately linked. In the final analysis, universalizability requires us to practice the reversibility of standpoints by extending this to the viewpoint of humanity. Such a capacity is essential to being a good partner in a moral conversation, and is itself furthered by the practice of moral conversation. (Benhabib 1992, 52)

Reversibility of perspectives makes the dialogue to be intraparadigmatic, a concept Benhabib uses. This is also a fruitful way to avoid populistic arguments. All included in the dialogue could analyze each argument and the premises behind, to see if the argument holds. Both aspects, the continuous process and the reversibility of perspectives, differ from the dialogue mentioned by Foster et al. (2020) in Sect. 3.5.2. In Foster et al. there is an idea of the dialogue as a presentation and a response, but that might imply a power relation that hinders those informed from reversing the perspectives as well as being able to disagree. However, reversing the perspectives and being able to disagree could further the process into more dialogue with the result of a deeper understanding of the complex problem. In this aspect, I am convinced that Benhabib's way of analyzing the dialogue form is far better than the suggestion by Foster et al. within the Planetary Health approach. Further, she specifies who to include:

> According to my formula, we are to imagine whether if I and all those whose actions would affect me, and by whose actions I would be affected, were to engage in a moral conversation, governed by the procedural constraints of universal respect and egalitarian reciprocity, we could adopt this as a principle of action. (Benhabib 1992, 37–38)

This quotation might be interpreted in an idealistic way, but Benhabib's framework is concrete and should take place in reality. Benhabib questions Rawls procedural framework of ethics known as the "veil of ignorance" as well as the wide reflective equilibrium on the fact that it is based on "anonymous others". In Benhabib's view an ethical framework needs to focus on *concrete others*. Her feminist and non-colonialist view is clear on the problem that idealized people is far from true people with all their complexity.

Would other than humans count? Benhabib extends the concept of rights outside the sphere of human rights. She claims that animals, at least those in our care, also should be attributed to moral rights, although they cannot have human rights (Benhabib 2014). Her view on how animals would influence the dialogue is not discussed.

She meets the critique that her view might be relativistic in the passage below where she tries to introduce how to justify different ways of living as better than others.

> Thus no way of life is prima facie superior to another and the prima facie validity it confers upon certain normative practices cannot be taken for granted if one cannot demonstrate with reasons to others who are not members of this way of life and even to skeptics among one's way of life as to why these practices are more just and fair than another. (Benhabib 1992, 42)[5]

[5] Prima facie is used in some ethical and legal discussions to indicate what seems clear to be done at first sight of a case, but not fully demanding if a deeper examination of the issues at stake shows another solution of the case. A prima facie view could be adjusted to circumstances and that adjustment could also influence forthcoming decisions.

Therefore, the dialogue will deal with these matters trying to establish good arguments for the choices of actions despite the differences in value ground regardless of cultural, ethical or religious origin.

7.3.1 *Dialogue Ethics: Bats Case*

In the Bats case, the ethical procedure is one ethical dialogue which is iterative and open until one has reached an ethical deliberation. There are no prima facie rules limiting the agenda. The process will work with understandings and misunderstandings as well as agreements and disagreements until one has reached the end of the ethical deliberation. Core is the concrete others. Facts on circumstances are therefore needed on human livelihoods as well as risks for disease transmission.

When choosing participants, Benhabib works for an inclusion of all human groups in society and aims for an analysis of power so that also marginal groups have their voice. In this case, if one takes the example of Covid-19 which origin has been connected to bats, the entire human community with all its different subgroups might be involved in the deliberation. One ethical example that could be discussed globally would be unfair distribution of vaccine (for two examples see Gayle and Childress 2021; Jecker 2021). All participants need to have the ability of reversibility of perspectives.

In the local livelihood where interactions with bats are part of the life one could include different interests such as agriculture (bat droppings, pollination), forest companies (cutting down woods), and Indigenous peoples (sacred places). Indigenous people's view needs also to be open for consideration in this process.

Bats might be included as moral objects, although Benhabib seems to prefer animals in human care as those being considered in the dialogue. Aspects concerning plants, biodiversity, and ecosystems only influence the dialogue indirectly because they seem to be outside those that have moral status in Benhabib's framework.

7.3.2 *Dialogue Ethics: Animal Assisted Therapy Case*

In the Animal Assisted Therapy case, the ethical procedure is one ethical dialogue which is iterative and open until one has reached an ethical deliberation. There are no prima facie rules limiting the agenda. The process will work with understandings and misunderstandings as well as agreements and disagreements until one has reached the end of the ethical deliberation.

When choosing participants, this case is fairly simple. Concrete others are humans involved in the therapy, the treater and the one receiving treatment, the horse, and others that could participate in therapy. All participants in the dialogue need to have the ability of reversibility of perspectives. Because Benhabib acknowledges rights (although not human) at least in animals in human care, her dialogue could encompass all directly involved. Aspects concerning plants, biodiversity, and ecosystems would fall outside the moral status in Benhabib's framework. Still, they could indirectly influence the discussions in the dialogue by being valuable either for humans or for horses.

7.3.3 Dialogue Ethics: Oceans Case

In the Oceans case, there are similarities with the Bats case. The ethical procedure is one ethical dialogue which is iterative and open until one has reached an ethical deliberation. There are no prima facie rules limiting the agenda. The process will work with understandings and misunderstandings as well as agreements and disagreements until one has reached the end of the ethical deliberation. Core is the concrete others. Facts on circumstances are therefore needed on both communities depending on oceans as part of their homes and other communities using resources from these communities.

When choosing participants, Benhabib works for an inclusion of all human groups in society and aims for an analysis of power so that also marginal groups have their voice. In this case, small island development states are particularly vulnerable due to risks of sea-level rising and should be included.

Aspects concerning marine animals, plants, biodiversity, and ecosystems would most probably only influence the dialogue indirectly because they seem to be outside those that have moral status in Benhabib's framework. However, there might be in some cultures wild animals that are in human care, like cormorants used for fishing, which then should be included.

7.3.4 Conclusion Dialogue Ethics

Although I appreciate the theory behind Benhabib's dialogue procedure, I have found it difficult to apply to the cases. There seem to be less to work with than expected. This might be the case due to the strong focus on humans. Also, a public deliberation might be very hard to manage due to the huge amount of people that might be involved, for example if one should discuss vaccine sharing for covid-19 globally. Is Benhabib's theory to open and inclusive? Maybe, but on the other hand we need as much perspectives as possible to be able to solve complex problems. I believe that this is still a suitable framework to use, but I will further develop the theory of dialogue trying to find a more suitable version beginning in the next section.

7.4 A Two-Stage Dialogue

Even if Seyla Benhabib's dialogue ethics is one of the most suitable that exist for One Health approaches, I am not fully satisfied. The open-ended inclusive style in the dialogue with a focus on the procedure is praiseworthy and I am not able to further develop that. What I would like to propose as a development concern what occurs prior to the dialogical deliberation. I believe one must go further into the initial part of the dialogue, the pre-framing part before the dialogue takes place.[6] There are several questions one must ask prior to the dialogue such as: Who or which are affected? How do you choose participants? Could one work with advocates? What issues should be discussed and how? How should these be discussed? For how long time? All these questions seem to be somewhat left out if one only has Benhabib's version of dialogue ethics. There are issues embedded in cultures and worldviews that might bias the dialogue if one starts with the assumption that only reversibility of perspectives and concrete others will do the work. The list of questions could be made considerably long if one has as an aim to cover different ethical, cultural and religious positions and bridge anthropocentrism, zoocentrism, biocentrism and ecocentrism. Most probably there needs to be a dialogue also about ethical forms. Therefore, I propose that dialogue ethics in this case must be developed to a two-stage dialogue.

7.4.1 First Stage of the Dialogue Process

The first stage of the dialogue process is a dialogue about creating the procedure for the second ethical dialogue. This is a pre-framing dialogue that for example decides answers to questions such as:

- Which core values should we agree on that we take for granted?
- Which stakeholders and which representation of stakeholders[7] is needed to make the dialogue inclusive and just?
- How to discuss the matters so that it is respectful?
- When have matters been deliberated enough and what outcome is intended?

---

[6] During the process of developing this aspect, I was reassured when in Summer 2023 preparing a talk on Vilhjálmur Árnasons' contribution to bioethics at a conference in Riga, Latvia. His paper on scientific citizenship in a democratic society emphazised in a clear way the importance of pre-framing before the dialogue gets started so that the dialogue will better facilitate for example policy writing (Árnason, 2012). During the conference we had further talks on pre-framing issues that furthered my thoughts into what I am now proposing.

[7] The question of advocates has grown within animal and environmental ethics. Could animals, plants, and ecosystems also gather around the table? The idea of advocates for animals has been raised by Nussbaum (2022) and she relies on argumentation made by Sue Donaldson and Will Kymlicka (2011). For trees and ecosystems, Christopher D. Stone has made an early contribution about the possibility of creating guardianship for those within environmental law (Stone, 1972).

Core values could for example be human rights, but that is open for discussion. Respect is to acknowledge relevant cultural and religious knowledge. That is a way of recognizing claims from Indigenous peoples (see Sect. 7.2.2.) as well as the NEOH suggestion by Laing et al. (2023; described in Sect. 3.7). To engage in respectful discussion one must consider how to best include groups that value individual knowledge alongside groups valuing collective knowledge. This will also influence power relations in the dialogue. Outcome could be a consensus view or deciding that two out of five solutions are ethically sound etc. Therefore, this dialogue is case-based and an ethically constructed procedural method.

Those participating in this first stage of dialogue must have ethical competencies in an inclusive and general way so that they are able to demarcate the boundaries for what to be discussed and which people that could have the competency to discuss the particular ethical issues (cf. the claims from Laing et al. 2023). The people making the choice of people in the first stage need to be people that is inclusive and fair. It could be difficulties gathering opposing parties around the table. Barriers could for example be linked to colonialism, religious disputes, gender issues etc. Here, the ethical idea of Ubuntu/Botho (a framework related to identity and personhood) evident in African indigenous languages (further analyzed in Sect. 7.4.2) could serve as a helping hand to gather also those that has opposing views. If all included has sufficient competencies, one can safeguard that different value systems and ethical systems might be present.

The way of creating a fair and just dialogue will take time in the beginning, especially when the dialogue needs to bridge barriers and make people on opposing sides to meet and discuss. Mutual trust takes time to build up and it is also something that could be ripped away if someone is decisive.[8] With time, this process could get faster, but a caution is in its place. Wicked problems, that One Health deals with (see Chap. 1), tend to have inherent obvious obstacles and hidden obstacles. To be able to solve the latter, those people who choose need to take their time in the process.

Benhabib thinks that these pre-framing issues might be solved within her dialogue method. She aims for a dialogue without prima facie rules. The point with my suggestion of a pre-framing dialogue is to bring these prima facie principles that we all have to the fore; openly to discuss them and whether they should be included. I think that without this first stage, one cannot even gather participants very far away from each other at the table. The way of including people with ethical competence to choose participants in the first stage might be questioned of being too paternalistic. Benhabib might argue that she has a non-paternalistic view focusing more on open public deliberation that also places claims of inclusion in the dialogue on the groups that want to participate. I think that these complex issues need to be addressed

[8] Trust is a complex concept that could refer to different things. Two of the most important in One Health approaches will be trust in institutions and trust in individuals. Onora O'Neill also makes a distinction of trust and trustworthiness which could be helpful. For a short introduction one can read her BBC Reith Lectures (O'Neill, 2002). For an interesting discussion between O'Neill and her critics Annette Baier, Karen Jones and Daniel Weinstock see Archard et al. (2013).

by people with different kinds of expert knowledge (where experts should be understood in a wide sense which includes all kinds of people suggested by the One Health approaches). This places me somewhere in between the suggestion of Foster et al. (which is more expert driven) and Benhabib.

7.4.2 *Choosing Participants Including the View of Ubuntu*

One important perspective on how to include stakeholders within a dialogue could be used from the philosophy of *ubuntu*, which has originated in African bantu-speaking peoples but nowadays regarded as a pan-African communitarian ideology (Eze 2023). Michael Onyebuchi Eze (2023) distinguishes two versions of ubuntu (also referred to as *botho*). One is an essentialist version, which according to him, has its drawback in being closed to change and only applicable within a specific African culture. The other is a performative version, with openness and sensitivity to historical change, internal variations in culture and therefore much more inclusive and transferable to other settings than its origin. Eze prefers the latter and I will also follow that version in this chapter. Central in ubuntu is the dialogue and a specific view on the individual and who to have a dialogue with.

> It is this nature of ubuntu as open-ended and discursive that underlies ubuntu as a philosophy of dialogue and human reconciliation. A person is a person through other people—this *other people* is a reference to those who are not members of my community, they could be anyone in their strangeness and oddities or as indicated earlier, animals and non-sentient beings (Eze 2023, 287–288, original emphasis)

The phrase "a person is a person through other people" claims that an individual is never isolated. Without interaction with others there is no individuality. This is similar to discussions of *the other* or *otherness* in the western continental philosophical tradition, but with an interesting difference that will be helpful here. Others could besides family and their own community, be others outside the community, others that I do not understand, and my enemies.[9] The idea that even our enemies should be included in the dialogue is worth further discussion, because that could be a solution to solve some cases where there are conflicts between stakeholders.

> Ubuntu dialogue is also an act of revolution, it is an act of rupture, interruption, and revolt against normality. Dialogue unmasks the other's strangeness and brings them to the realm of familiarity. Familiarity is an act of humanization. In interacting with others different from us, we get to know them. … We learn of our differences and sameness" (Eze 2023, 289)

In this quote one can grasp that ubuntu dialogue and the idea of individuality as always being a relation to others, links to the idea of human nature. Eze further specifies:

---

[9] This resembles a position held by Donna Haraway (2018, 105), "A livable world is remade with disregarded human persons and other displaced beings, or not at all. A livable world also requires making ontological room for beings that do not fit one's cast of characters."

> We are each other's creators and in this process of interaction and dialogue we realize our full humanity. Since we are co-creators of each other, a rule for dialogue is the understanding that the other person is still part of my humanity (Eze 2023, 289)

So, the enemy to me is also creating what is me. My human nature is constructed by all these different individuals that I meet, have relations with, have different feelings about and so forth. Will the inclusion of my enemies turn me to a moral saint[10] with the risk of being vulnerable to one's enemy? I believe that this is reciprocal. Including one's enemy into dialogue also puts pressure and demands on the enemy to follow the rules and commitments. That person needs also to be open to discussing my own version and why I see the other as an enemy as well as his own version. This view of human nature is closely linked to ethics. Eze quotes Desmond Tutu, the South African former archbishop:

> I am human because I belong. I participate, I share (Eze 2023, 289)

Is ubuntu only anthropocentric? No, as already quoted above, Eze mention that ubuntu also can include animals, non-sentient beings, those that have passed away, and gods and goddesses. This makes the ubuntu view of dialogue extremely inclusive in relation to culture, religion as well as non-human individuals. Therefore, ubuntu might be useful as one of the perspectives to include in a future road map of ethics for One Health approaches. In the process of including stakeholders, ubuntu might be one guide for an inclusive procedure-based One Health dialogue.

7.4.3 Second Stage of the Dialogue Process

After the first stage of dialogue, then there will be a second stage of dialogue where the ethical matters concerning the case are discussed. Those chosen for the second stage would be gathered. Here ethical competency needs to be more specific and context specific.[11] Some of these are there for being stakeholders for their own group, others are there for being advocates for animals, plants, ecosystems, or those that cannot participate in one way or another.[12] Those involved in the first dialogue and the second could therefore be different. Those considered for the first dialogue need a broad ethical competency to bridge the barriers and identify those that have specialist ethical competence in the issues at stake (which differs depending on matters). However, if a person has competency for both stages, that person could be a part of both dialogues. It is in this case the particular competency in both aspects that is guiding who to include.

---

[10] Susan Wolf (1982) introduced the problem of being a moral saint and the risk of doing "to much" in ethics compared to others.

[11] Compare this to the ethical competency described by Laing et al. (2023, see Sect. 3.7). In my suggestion one must be aware that ethical competency can be at least in two kind, which both are valued and needed for the process.

[12] See footnote 7.

This second dialogue resembles the dialogue favored by Benhabib (1992) where one chooses among ethical alternatives and tries to find the best ethical motivation. That dialogue needs to be formed by concrete others, an iterative process until enough depth has been reached and all solutions deliberated. The process must follow the rules established in the first dialogue and be fair in allowing the reversibility of perspectives.[13]

To justify the entire procedure, one must ask whether all could agree that the steps have been followed sufficiently. Then the outcome is justified.

A way to summarize this two-stage dialogue is in this list of steps:

Stage 1

1. One must agree that ethics is an agreed ethical constructivism.
2. One must agree on specific core values that are the foundation of the One Health approaches. These values (at this point although not specified particularly what they mean) could for example be health, justice, transdisciplinism, and that one should allow all four perspectives I have discussed (anthropocentrism, zoocentrism, biocentrism, and ecocentrism).
3. One must agree on a procedure to find relevant stakeholders and aspects so that one can, for example, choose the right representatives, the right claims etc.
4. One must agree on a procedure on how to discuss matters, when a rich enough deliberation has been made and what kind of outcome one would aim for.

Stage 2

5. In this procedure, several aspects of ethics are specified. These could be how to define certain values, which principles or intuitions that are helpful, which arguments that are valid, how to weigh between different possible ethical solutions.
6. Finally, a deliberated ethical decision could be found that is justified through the process.

7.4.4 *Critical Remarks*

The main objection might be that this idea tends to open for relativism in ethics. I will here echo Benhabib's thoughts that relativism is not problematic in this case. By using ethical competency as a guiding principle, one might also diminish the risk of failure in the procedure.

This is also a very complex method of ethics, maybe one of the most complex. But complex cases, involving different worldviews, valuations, ethical systems, different species and levels of system (individual, population or ecosystem) need a complex ethical system in order not to lose important aspects. This is not exclusive

[13] As I wrote in Sect. 7.3, the proposal on dialogue in Planetary Health (Sect. 3.5.2) made by Foster et al. (2020), lack these aspects and tend to be a one-sided dialogue where the power is in the hands of the person with the knowledge on Planetary Health matters.

for ethics. Complex models that are elongated in time have already been suggested within One Health approaches for knowledge sharing (Binot et al. 2015).[14]

My suggestion has also gained strength from the suggestions from Indigenous Peoples claim for a procedure-oriented ethics where all is included from the start like the Akwé: Kon guidelines (Secretariat of the Convention on Biological Diversity 2004) mentioned in Sect. 7.2.2.

The initial steps and especially the *second step* in the first stage of the two-stage dialogue (Sect. 7.4.3) differ from many of the frameworks presented in this book, because it tries to avoid hidden assumptions. As an example, many states consider themselves democratic and democracy is a positively value-laden term. However, bringing statesmen from these different states together discussing what democracy is will bring forward different conceptions that also might be contrary to each other. In ethics for One Health approaches with its diverse views, these discussions are needed to settle the agenda for ethics.

In the *third step* of the first stage, the procedure to find relevant stakeholders could consist of several puzzle pieces. I suggest one procedure might be the ubuntu inclusion of people in the dialogue described in Sect. 7.4.2 (Eze 2023) to give respect to different cultures and religions as well as not opting enemies to the process out from the procedure.

How could animals, plants, or ecosystems take part in the second stage? In the third step of the first stage is an option to think about advocates or representatives. People that have particular knowledge about species-specific aspects, ecosystem functioning, or other aspects needed to safeguard animals, plants, or ecosystems in the dialogue procedure.

7.4.5 Two-Stage Dialogue: Bats Case

The Bats case consists of humans and bats in their environment. When this case is analyzed with the framework of the Two-stage dialogue, human, bats, and their environment is analyzed.

Stage 1

In the first stage one needs to decide on how to perform the dialogue and what is at stake. Obvious in this case is the need to bridge western science with Indigenous knowledge. Therefore, the participants in this first dialogue must be able to understand the different value systems and ethical theories that might have implications on these matters. Participants from western science and Indigenous peoples must be included and they must be open that the process of choice between the groups as well as expected timeframes might differ according to cultural and/or religious matters. For example, the Indigenous people should choose their participant according

[14] Also, Binot et al. (2015) put some emphasis to pre-framing. However, their method and aim differ considerably from my suggestion.

to their cultural way of choosing representatives. The timeframes must respect the participating different cultures.

All affected need to be included in one way or another. One can think about representatives or advocates for bats and the environment. These need to have knowledge on bat species, ecosystem services, food web dynamics of the ecosystems, ecosystem, biodiversity, disease transmission risks, etc.

Stage 2

Stage 2 has similarities with Seyla Benhabib's dialogue process. The stage 2 ethical dialogue is iterative and open until one has reached an ethical deliberation but one needs to follow the procedure developed in stage 1. One could in Stage 2 use and discuss ethical concepts, principles, theories, or frameworks to reach an ethical deliberation. Besides including Aldo Leopold's Land ethic (which I think suitable, see Sect. 6.3), others might argue for other environmental ethics positions as well as animal ethics positions.

7.4.6 Two-Stage Dialogue: Animal Assisted Therapy Case

The Animal Assisted Therapy (AAT) case consists of humans and animals in our care. When this case is analyzed with the framework of the Two-stage dialogue, humans and animals involved in the therapy is analyzed.

Stage 1

In the first stage one needs to decide on how to perform the dialogue and what is at stake. This case is somewhat narrower than the other two cases and decisions might be easier to reach.

All affected need to be included in one way or another. One can think about representatives or advocates. Concrete others are humans involved in the therapy, the treater and the one receiving treatment, the horse, and others that could participate in therapy. One must not forget to include persons that could speak for the horse due to their established knowledge on horse behavior, health and welfare. Indirectly, the environment, other animals as well as other people might be afflicted in some way or another by therapy.

One important value question to discuss is the value of animals. An outlook on animal legislation shows that many nations regard their animals as things, something one can own or possess. Some countries have initiated and partly finalized a change in values within their legislation from instrumental to intrinsic, giving animals protection that could be equated with rights although they often are not as strong as human rights (Lerner 2008; Fasel and Butler 2023).

Stage 2

Stage 2 has similarities with Seyla Benhabib's dialogue process. The stage 2 ethical dialogue is iterative and open until one has reached an ethical deliberation but one needs to follow the procedure developed in stage 1. One could in Stage 2 one could use and discuss ethical concepts, principles, theories, or frameworks to reach an ethical deliberation. Besides including Martha Nussbaum's Capability approach (which I think suitable, see Sect. 5.5), others might argue for an inclusion of the Four Principles of animal ethics (see Sect. 5.6) or other animal welfare perspectives of different kinds.

7.4.7 Two-Stage Dialogue: Oceans Case

The Oceans case consists of humans and their relation to oceans. When this case is analyzed with the framework of the Two-stage dialogue, several aspects of that relation is analyzed.

Stage 1

In the first stage one needs to decide on how to perform the dialogue and what is at stake. Obvious in this case is the need to bridge western science with Indigenous knowledge. Participants from these groups must be included and one must be open to the fact that the process of choice between the groups might differ according to cultural and/or religious matters. For example, the Indigenous people should choose their participant according to their cultural way of choosing representatives.

Besides knowledge in different kinds of ethics, other kinds of knowledge important for the process would be related to the issues of climate change, avoidance of pollution, food webs in the ocean, biodiversity, sustainable living etc.

All affected need to be included in one way or another. One can think about representatives or advocates. Regarding this case and sea level rising one should particularly include the small island developing states. Also, fishing companies, regulatory bodies, climate scientists, and ecologists need to be invited.

Stage 2

Stage 2 has similarities with Seyla Benhabib's dialogue process. The stage 2 ethical dialogue is iterative and open until one has reached an ethical deliberation but one needs to follow the procedure developed in stage 1. One could in Stage 2 one could use and discuss ethical concepts, principles, theories, or frameworks to reach an ethical deliberation. Besides including Aldo Leopold's Land ethic (which I think suitable, see Sect. 6.3), others might argue for other environmental ethics positions as well as animal ethics positions.

7.4.8 Conclusion Two-Stage Dialogue

This framework has the strength to include different cultural and religious perspectives and might also acknowledge and incorporate other ethical frameworks that might be fruitful in the second stage of the procedure. For the Animal Assisted Therapy case, the first stage might be superfluous, but it might be performed in a simpler and shorter form to be sure not to leave anything out. The first stage in the Bats and Oceans case might be challenging to overview and could be prolonged in time to find a sufficient group of people that could participate in the second dialogue. But as I already have stated, complex problems might require complex solutions to get acceptance from all sides on decided solutions.

7.5 Summary

Within this chapter pluralistic ethical frameworks have been introduced and discussed. The protection of Indigenous peoples and local communities' knowledge by the Convention of Biological Diversity implies that Indigenous peoples and local communities cannot be avoided in the process of dialogue. Their voices need to be heard.

Seyla Benhabib's framework of ethical dialogue meets demands on intersectionality from Chap. 3, especially on cultural and religious inclusion, non-colonial perspectives, and gender inclusiveness.

I criticize Benhabib for giving too little emphasis on the pre-framing part of the dialogue by proposing a more developed version of dialogue ethics. This is a two-stage dialogue where the first stage is a pre-framing dialogue where the form of dialogue could be decided and participants in the second stage chosen. Here, the framework of Ubuntu might be valuable for how to choose participants for dialogue. One should consider all persons influencing oneself, also those that are enemies to you. The second stage of dialogue is similar to the dialogue process that Benhabib suggests.

Next chapter will summarize the findings of all chapters and normatively suggest a roadmap forward towards an ethical framework for One Health approaches.

References

Archard, David, Nonique Deveaux, Neil Manson, and Daniel Weinstock, eds. 2013. *Reading Onora O'Neill*. Oxon and New York: Routledge.

Árnason, Vilhjálmur. 2012. Scientific citizenship in a democratic society. *Public Understanding of Science* 22 (8): 927–940.

Benhabib, Seyla. 1992. *Situating the self: Gender, community and postmodernism in contemporary ethics*. Cambridge and Malden, MA: Polity Press.

Benhabib, Seyla. 1996a. The democratic moment and the problem of difference. In *Democracy and difference: Contesting the boundaries of the political*, ed. Seyla Benhabib, 3–18. Princeton, NJ: Princeton University Press.

Benhabib, Seyla. 1996b. Toward a deliberative model of democratic legitimacy. In *Democracy and difference: Contesting the boundaries of the political*, ed. Seyla Benhabib, 67–94. Princeton, NJ: Princeton University Press.

Benhabib, Seyla. 2014. Defending a cosmopolitanism without illusions. Reply to my critics. *Critical Review of International Social and Political Philosophy* 17 (6): 697–715.

Binot, Aurelie, Raphaël Duboz, Panomsak Promburom, Waraphon Phimpraphai, Julien Cappelle, Claire Lajaunie, Flavie Luce Goutard, Tanu Pinyopummintr, Muriel Figuié, and François Louis Roger. 2015. A framework to promote collective action within the One Health community of practice: Using participatory modelling to enable interdisciplinary, cross-sectoral and multi-level integration. *One Health* 1:44–48.

Brant, J., S. Stagg Peterson, and N. Friedrich. 2023. Partnership research with Indigenous communities: Fostering community engagement and relational accountability. *Brock Education Journal* 32 (1): 99–118.

Donaldson, Sue, and Will Kymlicka. 2011. *Zoopolis: A political theory of animal rights*. Oxford: Oxford University Press.

Eze, Michael Onyebuchi. 2023. Ubuntu/Botho. In *Key concepts in world philosophies*, ed. Sarah Flavel and Chiara Robbiano, 283–291. London: Bloomsbury Academic.

Fasel, Raffael, and Sean C. Butler. 2023. *Animal rights law*. Oxford: Hart Publishing.

Foster, Alexander, Jennifer Cole, Ivica Petrikova, Andrew Farlow, and Howard Frumkin. 2020. Planetary health ethics. In *Planetary health: Protecting nature to protect ourselves*, ed. Samuel Myers and Howard Frumkin. Washington: Island Press.

Gayle, Helene D., and James F. Childress. 2021. Race, racism, and structural injustice: Equitable allocation and distribution of vaccines for the covid-19. *The American Journal of Bioethics* 21 (3): 4–7.

Haraway, Donna. 2018. Staying with the trouble for multispecies environmental justice. *Dialogues in Human Geography* 8 (1): 102–105.

Herman, R. D. K. 2014. Approaching research in Indigenous settings: Nine guidelines. In Toolbox on the research principles in and Aboriginal context: Ethics, respect, equity, reciprocity, collaboration and culture (pp. 1–11). FNQL HSSC, UQATn CRDP, and DIALOG Network. https://reseaudialog.ca/wp-content/uploads/2021/01/Toolbox_Research_Principles_Aboriginal_Context_eng.pdf

Jecker, Nancy S. 2021. Global sharing of covid-19 vaccines: A duty of justice, not charity. *Developing World Bioethics* 23:5–14.

Laing, Gabrielle, Eleanor Duffy, Neil Anderson, Nicolas Antoine-Moussiaux, Maurizio Aragrande, Caetano Luiz Beber, John Berezowski, Elena Boriani, Massimo Canali, Luis Pedro Carmo, Ilias Chantziaras, Glen Cousquer, Daniele De Meneghi, Ana Gloria Rodrigues Sanches da Fonseca, Julie Garnier, Martin Hitziger, Thomas Jaenisch, Hans Keune, Claire Lajaunie, Lorena Franco Martinez, Rebecca Maudling, Marie K. McIntyre, Barry J. McMahon, Alberto Munoz Prieto, Liza Rosenbaum Nielsen, Ranya Özçelik, John W.A. Rossen, Simon R. Rüegg, Sara Savić, Margarida Pires Simoes, Deborah J. Thomson, Laura Tomassone, Asta Tvarijonaviciute, Manuela Vilhena, Barbara Vogler, and Barbara Häsler. 2023. Advancing one health: Updated core competencies. *CABI One Health*. https://doi.org/10.1079/cabionehealth.2023.0002.

Lerner, Henrik. 2008. The concepts of health, well-being and welfare as applied to animals. A philosophical analysis of the concepts with regard to the differences between animals. Ph.D. Thesis. Linköping: Department of Medical and Health Sciences, Division of Health and Society, Linköping University. Linköping Studies in Arts and Science No. 438.

Manguvo, Angellar, and Benford Mafuvadze. 2015. The impact of traditional and religious practices on the spread of Ebola in West Africa: Time for a strategic shift. *PanAfrican Medical Journal* 22 (Suppl. 1): 9.

Marshall, Katherine. 2017. Roles of religious actors in the West African Ebola response. *Development in Practice* 27 (5): 622–633.

Middleton, Jo, Jackie A. Cassell, Gavin Colthart, Francesca Dem, James Fairhead, Michael G. Head, Joao Inacio, Mavis Jimbudo, Moses Laman, Vojtech Novotny, Mika Peck, Jonah Philip, William Pomat, Shen Sui, Peter West-Oram, and Alan Stewart. 2020. Rationale, experience and ethical considerations underpinning integrated actions to further global goals for health and land biodiversity in Papua New Guinea. *Sustainability Science* 15:1653–1664.

Nussbaum, Martha C. 2022. *Justice for animals: Our collective responsibility*. New York: Simon & Schuster.

O'Neill, Onora. 1996. *Towards Justice and virtue: A constructive account of practical reasoning*. New York and Melbourne: Cambridge University Press.

O'Neill, Onora. 2002. *A question of trust: The BBC Reith lectures 2002*. Cambridge: Cambridge University Press.

O'Neill, Onora. 2013. Responses. In *Reading Onora O'Neill*, ed. David Archard, Monique Deveaux, Neil Manson, and Daniel Weinstock. London and New York: Routledge.

Secretariat of the Convention on Biological Diversity. 2004. Akwé: Kon voluntary guidelines for the conduct of cultural, environmental and social impact assessment regarding developments proposed to take place on, or which are likely yo impact on, sacred sites and on lands and waters traditionally occupied or used by indigenous and local communities. Montreal, 25pp. (CBD Guidelines Series).

Secretariat of the Convention on Biological Diversity. 2011. Convention on biological diversity: Text and annexes. (First adopted 22 May 1992). Montreal, 30 pages. https://www.cbd.int/doc/legal/cbd-en.pdf

Stone, Christopher D. 1972. Should trees have standing? Toward legal rights for natural objects. *Southern California Law Review* 45:450–501.

Tunón, Håkan, Marie Kvarnström, and Henrik Lerner. 2016. Ethical codes of conduct for research related to Indigenous peoples and local communities—Core principles, challenges and opportunities. In *Ethics in Indigenous research: Past experiences – future challenges*, ed. Anna-Lill. Drugge, 57–81. Umeå: Vaartoe – Centre for Sami Research. http://umu.diva-portal.org/smash/get/diva2:943266/FULLTEXT03.pdf

Warren, Karen J. 2015. Feminist Environmental Philosophy. In: Edward N. Zalta (Ed.) The Stanford Encyclopedia of Philosophy (Summer 2015 Edition), URL = https://plato.stanford.edu/archives/sum2015/entries/feminism-environmental/. Retrieved 27 Dec 2023.

Wolf, Susan. 1982. Moral saints. *Journal of Philosophy* 79 (8): 419–439.

Part III
Roadmap Ahead

Chapter 8
A Roadmap for Future Directions

> *Reality is always harder to master than visions. But we always must have faith that visions can be transferred down to reality. (Edberg, 1993, 127) (My translation from Swedish. Original quote "Verkligheten är alltid mer svårbemästrad än visionerna. Men vi måste alltid vägledas av tron att visioner kan flyttas ner i verkligheten".)*

Abstract This chapter concludes the analysis throughout the book and provides a roadmap with suggestions for further application of ethics in One Health approaches. First, I make a synthesis conclusion of the book. I then turn to some practical recommendations for research, policy making and education in One Health approaches. Finally, I will draw the roadmap for ethics in One Health approaches with a summary of which frameworks that seem to be fruitful and which frameworks that can be avoided. My main finding in the book is that there is a need for a two-stage dialogue within One Health approaches and that might be the most sufficient solution. Other fruitful frameworks are the Capability Approach and the Land Ethic Decision Model.

Keywords Education · Policy · Research · Recommendations · Intergenerational ethics · Culture · Gender · Religion · Intersectionality

H. Lerner, *Ethics for One Health Approaches: A Roadmap for Future Directions*, Advancing Global Bioethics 22,
https://doi.org/10.1007/978-3-032-05239-1_8

8.1 Introduction

Finally, I have reached the point where I will try to spell out what I believe would be the most promising way forward in choosing roads in the roadmap for One Health ethics.[1] Several roads seem to be promising, and time will tell which ones that will be heavily used. As is the case with a roadmap, I will here indicate major roads that make it easy to reach the goal in an ethically sound way. Other roads that at the moment seem to be too winding or too narrow for the One Health approaches will be left out but could by development seem to be more promising in the future. I hope it will encourage others to start a dialogue where this proposal will be criticized, better articulated, expanded, or exchanged for a better solution.

In this chapter, I will first make a synthesis conclusion of the book, then I will give recommendations for science, policy making, and education. Finally, I draw the roadmap and its conclusions. As Rolf Edberg (1993) states in the epigraph above, vision and policy must go hand in hand. A vision for One Health ethics needs to be able to meet the requirements of the field as well as being applicable to real world problems in order to reach guidelines for everyday use.

The three cases I have used as a test for the ethical frameworks, the Bats case, the Animal Assisted Therapy case, and the Oceans case, are only three of many cases that we must face now and in the future. New cases will also certainly arise that need to be handled in the future. One Health ethics needs to be a down-to-earth science based on applied ethics and relating to empirical matters to be sufficient and fruitful. I will now turn to a summary of what has been learned through the chapters of the book.

8.2 Synthesis of the Book

Chapter 1 introduced the field of One Health ethics, which is a fairly new topic in bioethics. Due to the scope of One Health approaches finding an ethical framework for these approaches might be one of the widest bioethics possible. I also provided some thoughts on my view of ethics. I take a rather pragmatic view, thinking that more than one ethical framework might be useful, but each proposed framework needs to be in line with the core values within the approaches. I also presented four evolving themes in ethics from the twentieth century up to the present day that one needs to consider in order to have a modern ethics. These were:

[1] This roadmap simile has a similar function as Onora O'Neill's (1996) simile of a blueprint which is mentioned in Chap. 7. My roadmap will, as her blueprint, specify some fundamentals and leave others out. As I mentioned in footnote 6, Sect. 1.2, "Roadmap" as a concept has been used earlier in One Health literature. Nicole de Paula (2021) has used it as giving the direction for Planetary Health in the future. Zohar Lederman (2018) has used it in a pragmatic way to avoid defending his claims of theory in his ethical thesis on One Health and culling (see Chap. 3). As is evident in this chapter, my intention with the concept differs from his.

1. Expanding ethics
2. Questioning the individualistic take
3. Pluralism
4. Constructivism

These have been guiding throughout the book. I have at a few places in this synthesis below indicated in brackets for the reader where these trends have had a rather strong influence on my argumentation. The three cases I mentioned in the last section were presented at the end of Chap. 1.

In Chap. 2, I analyzed the differences and similarities between four influential One Health approaches – One Health, EcoHealth, Planetary Health, and One Welfare – regarding scientific scope and core values. Several of the One Health approaches recognize that humans, animals, plants, and ecosystems have an inextricable connection, and health should be valued for all these. Planetary Health was considered anthropocentric, One Health and One Welfare zoocentric, and EcoHealth biocentric or ecocentric. This implies that an ethical framework for One Health approaches needs to bridge anthropocentric, zoocentric, biocentric, and ecocentric frameworks (Trend 1). A quite impossible task. Also, regardless of the different approaches, ethics needs to be a core aspect of all these approaches (including new ones arising). The ethical framework also needs to be inclusive in cultural and religious values due to the fact that at least One Health, EcoHealth, and Planetary Health acknowledge Indigenous knowledge, besides Western science, as knowledge base within the approach (Trend 3).

In Chap. 3, an overview of suggested ethical frameworks proposed by scientists within the One Health approaches clarified that there is still a call for ethics, and ethical discussion is a pristine field. However, as evident in Table 3.1, sufficient suggestions for ethical frameworks have been proposed to make an in-depth analysis. These frameworks were Principlism, Biophilia, Capability Approach, Four principles of animal ethics, Land ethic, as well as calls for ethics for care and ecofeminism, recognition of indigenous values, and ethical pluralism, which were analyzed in Chap. 4 to 7 and applied to the cases presented in Chap. 1.

Chapter 4 analyzed anthropocentric frameworks proposals, namely Principlism and Biophilia. None of these were considered fruitful for a working ethical framework. Naturally, this follows from the inherent claim argued for in Chaps. 2 and 3 that One Health approaches need an ethic that is not anthropocentric. Although anthropocentric frameworks might work for Planetary Health, it's too narrow to be used in all One Health approaches.

In Chap. 5, zoocentric frameworks mainly proposed for the One Welfare approach—the Capability approach and the Four principles of animal ethics—were analyzed. The capability approach seemed fruitful, at least for the AAT case (i.e. relations between humans and animals). The rich way of analyzing human and animal prospering could be guiding to safeguard a good life. The Four principles of animal ethics framework were however disregarded due to the problem of all principlism, that principle-based frameworks tend to end up in only weighing one particular value, in this case harms.

Chapter 6 considered biocentric and ecocentric frameworks and analyzed relational values as something to be added to the distinction between intrinsic and instrumental values. The main framework analyzed was Aldo Leopold's Land ethic where I developed a Land Ethic Decision Model so that the framework could more easily be applied to the cases. I showed how this framework could be fruitful for One Health approaches, especially for the relation between humans and ecosystems.

Chapter 7 was dedicated to the claim that pluralism in ethics is needed and that diversity in cultural and religious values needs to be part of an ethical framework (Trend 3). My main analysis was on Seyla Benhabib's dialogue ethics which I consider fruitful. However, I think that her framework could be further developed by Ubuntu, which is a theory downplaying the individual arguing that each individual is co-constructed by all others around them, including their enemies (Trend 2). To include enemies at the discussion table on the same demanding grounds as friends for a fair and just dialogue could be an opportunity to avoid polarization and find holistic solutions. My own suggestion of having a two-stage dialogue instead of the ordinary one facilitates this.

The point with a two-stage dialogue where two stages of dialogue is used facilitates intercultural and interreligious dialogue better than Benhabib's version. In the first dialogue participants pre-frame the ethical dialogue so that as many as possible could gather around the table to discuss. Here ethical competency is important to gather all perspectives. The first dialogue sets the agenda for how the second ethical dialogue will work (Trend 4). It also chooses participants for the second dialogue, which then will be an open-ended dialogue with similarities to Benhabib's version of dialogue ethics. The result of the second dialogue is ethical solutions to the case and if the procedure has been fair in both dialogues, the procedure is ethically good.

8.3 Recommendations for Researchers, Policy Makers, and Educators

The intention of this book is to establish how important ethics is for One Health approaches and that ethics cannot be omitted within the approaches any longer. In this section I will stress the importance of including ethics in research, policy making and education that follows from this study of ethics within the approaches.

8.3.1 Research

Ethics can no longer only be called for; it must be a core science within all inter- and transdisciplinary One Health approaches. In forthcoming research projects ethics must be as inherent in the inter- and transdisciplinary process as it is in human healthcare. Researchers must be fully aware of this and experts in ethics should be appointed to research groups. In the future umbrella depictions (like the ones in Lerner and Berg, 2015; García Pinillos et al., 2016; García Pinillos, 2018, see also Sect. 2.7.1), ethics should be added to the core sciences at the top of the umbrellas.

There is also a need for specific fields in applied ethics and bioethics for development of ethical frameworks within the One Health approaches. As evident in this book, many of the problems that occur in the world today is holistic and therefore particularly hard to solve and as I wrote in Chap. 1, One Health approaches might be the biggest challenge for bioethics. The imperative to bridge anthro-, zoo-, bio- and ecocentrism is out there to challenge. Is such a bridge possible although it now seems hard? This is also a field where one could be able to bridge philosophical strands from different philosophical traditions as well as from religious ethics due to its inter- and transdisciplinary nature.

8.3.2 *Policy Making*

For policy makers the importance of ethical competency (Sect. 3.7; Laing et al. 2023) is core. To be able to succeed in creating an ethically sound policy on implementation that could impact areas of the world better or worse, one must pinpoint and include those different people that have ethical competence. Also, one must strive for a diversity of perspectives along all these aspects of humankind that we include in diversity issues.

The strong emphasis in this book on procedural matters and the downplay of ethics based on principles implies that policy makers first need to establish an ethically sound procedure which all affected can agree on. To gather as much as possible, one needs to think through not just the ethical deliberation but also the pre-framing of the ethical deliberation. An agreed fair procedure could be open, and several ethical frameworks could be available for consideration. This is also reflected in the view on ethics as providing guidelines rather than complete instructions (O'Neill 1996, see also Sect. 7.2.1). This book has proposed that for example an ethical dialogue could work with either a Capability Approach or a Land Ethic Decision Model framework, or maybe both.

The roadmap suggested is also in line with intergovernmental bodies' work such as the United Nations work with human rights and sustainable development goals, and the Intergovernmental Platform on Biodiversity and Ecosystem Services work which in its latest version called Nature Futures Framework.

8.3.3 *Education*

As important as in human medicine (where public health ethics, ethics of care and medical ethics are taught), and veterinary medicine (where veterinary ethics and animal ethics are taught), ethics must be a subject in educating personnel that works within One Health approaches facilitating policy implementation. This has earlier been called for in studies of education within the approaches (Abed et al., 2021), but I believe that this book has clarified the need. At undergraduate level one must have at least an introduction in applied ethics together with work on cases. Also, students need to be encouraged to include ethical perspectives in their theses at different educational levels.

8.4 Roadmap for Further Research

So, as a final summary at the end of the book, how is this roadmap I am picturing situated? First, ethics within One Health approaches needs to be inclusive to intergenerational issues and consider humans, animals, plants, and ecosystems as all having moral status in one way or the other. It needs also to be inclusive concerning gender, cultural and religious perspectives often gathered under the umbrella of intersectionality. I have tried to establish that the future road to a functioning ethics for One Health approaches must be procedural and pluralistic. A main road might be my suggestion of a two-stage dialogue (Sect. 7.4) that would solve some of the problems of gathering a sufficient number of perspectives around the table. Then, one might solve those complex problems mentioned by David Waltner-Toews (2017) in the epigraph of Chap. 1. An alternative road might be the dialogue form argued for by Seyla Benhabib (Sect. 7.3), although with the risk of missing the targets. Other main roads seem to be the Capability Approach (Sect. 5.5) and the Land ethic (Sect. 6.3).

Maybe these three roads are connected. The two-stage dialogue sets the framework. The first stage dialogue is to understand the case or issue, setting the agenda and choosing participants. Ethical frameworks included in the second dialogue could for example be Martha Nussbaum's Capability approach, Aldo Leopold's Land ethic as well as other frameworks not yet analyzed. The Capability approach is particularly useful in issues concerning the well-being of humans and sentient animals. I believe this is one of the richest theories to grasp the full complexity of a sentient being as well as it is species-specific avoiding anthropomorphism. The Land Ethic Decision Model is particularly suited to dealing with complex issues concerning environmental aspects. It focuses on relations between humans and the rich variety in the environment and ecosystems. As I stated already in Sect. 1.2, I think that a possible way forward is to think about ethical theories working complimentary to each other. This is not an easy way to go because one often inherits problems from each of the theories in the combined theory, but I have doubts that we are able to reach a point where everyone with ethical competence agrees on *the* theory of ethics to work, which will fit every case. Further development of the roadmap to find a finer grained network of roads as well as existing alternative roads might be welcomed here.

Roads to avoid seem to be various versions of principlism (see Sects. 4.3 and 5.6), either they are anthropocentric or zoocentric. They seem to be non-inclusive.

Biophilia (Sect. 4.4) is also avoided because the framework lacks the possibility to be an ethical framework for decision making.

There must also be white spots on the roadmap, roads not known at the moment which could be explored. These are left for us to further think about and find. The field of One Health ethics is out there waiting for us.

References

Abed, Yehia, Madhumita Dobe, Eliudi Eliakimu, Rusmir Goletic, Tomiko Hokama, Ulrich Laaser, George Lueddeke, Linda Mans, Veronica Ormea, Monalisha Sahu, and Richard Seifman. 2021. The global one health environment: A learning module edited by 1HOPE-GPA (Governance, Policy, Advocacy). *South Eastern European Journal of Public Health*, Special Volume No. 1:1–205.

de Paula, Nicole. 2021. *Breaking the silos for planetary health: A roadmap for a resilient post-pandemic world*. Singapore: Palgrave Macmillan.

Edberg, Rolf. 1993. Vilopunkter. Wiken förlag. ISBN 91-7119-556-4. [In Swedish].

García Pinillos, Rebeca. 2018. *One welfare: A framework to improve animal welfare and human well-being*. Wallingford: CABI.

García Pinillos, Rebeca, M. Appleby, X. Manteca, F. Scott-Park, C. Smith, and A. Velarde. 2016. One welfare – a platform for improving human and animal welfare. Veterinsary Record, https://doi.org/10.1136/vr.i5470.

Laing, Gabrielle, Eleanor Duffy, Neil Anderson, Nicolas Antoine-Moussiaux, Maurizio Aragrande, Caetano Luiz Beber, John Berezowski, Elena Boriani, Massimo Canali, Luis Pedro Carmo, Ilias Chantziaras, Glen Cousquer, Daniele De Meneghi, Ana Gloria Rodrigues Sanches da Fonseca, Julie Garnier, Martin Hitziger, Thomas Jaenisch, Hans Keune, Claire Lajaunie, Lorena Franco Martinez, Rebecca Maudling, Marie K. McIntyre, Barry J. McMahon, Alberto Munoz Prieto, Liza Rosenbaum Nielsen, Ranya Özçelik, John W.A. Rossen, Simon R. Rüegg, Sara Savić, Margarida Pires Simoes, Deborah J. Thomson, Laura Tomassone, Asta Tvarijonaviciute, Manuela Vilhena, Barbara Vogler, and Barbara Häsler. 2023. Advancing one health: Updated core competencies. CABI One Health. https://doi.org/10.1079/cabionehealth.2023.0002

Lederman, Zohar. 2018. Stamping out culling: The ethics of culling from a one health perspective. Ph. D. thesis. Centre for Biomedical Ethics, National University of Singapore.

Lerner, Henrik, and Charlotte Berg. 2015. The concept of health in One Health and some practical implications for research and education: What is One Health? *Infection Ecology & Epidemiology* 5:25300. https://doi.org/10.3402/iee.v5.25300.

O'Neill, Onora. 1996. *Towards justice and virtue: A constructive account of practical reasoning*. New York and Melbourne: Cambridge University Press.

Waltner-Toews, David. 2017. Zoonoses, one health and complexity: Wicked problems and constructive conflict. *Philosophical Transactions of the Royal Society B* 372:20160171.

Index

H. Lerner, *Ethics for One Health Approaches: A Roadmap for Future Directions*, Advancing Global Bioethics 22,
https://doi.org/10.1007/978-3-032-05239-1

9783032052384